Patrick Moore's Practical Astronomy Series

Other Titles in this Series

Telescopes and Techniques (2nd Edn.)
Chris Kitchin

The Art and Science of CCD Astronomy
David Ratledge (Ed.)

The Observer's Year (2nd Edn.)
Patrick Moore

Seeing Stars
Chris Kitchin and Robert W. Forrest

Photo-guide to the Constellations
Chris Kitchin

The Sun in Eclipse
Michael Maunder and Patrick Moore

Software and Data for Practical Astronomers
David Ratledge

Amateur Telescope Making
Stephen F. Tonkin (Ed.)

Observing Meteors, Comets, Supernovae and other Transient Phenomena
Neil Bone

Astronomical Equipment for Amateurs
Martin Mobberley

Transit: When Planets Cross the Sun
Michael Maunder and Patrick Moore

Practical Astrophotography
Jeffrey R. Charles

Observing the Moon
Peter T. Wlasuk

Deep-Sky Observing
Steven R. Coe

AstroFAQs
Stephen Tonkin

The Deep-Sky Observer's Year
Grant Privett and Paul Parsons

Field Guide to the Deep Sky Objects
Mike Inglis

Choosing and Using a Schmidt-Cassegrain Telescope
Rod Mollise

Astronomy with Small Telescopes
Stephen F. Tonkin (Ed.)

Solar Observing Techniques
Chris Kitchin

Light Pollution
Bob Mizon

Using the Meade ETX
Mike Weasner

Practical Amateur Spectroscopy
Stephen F. Tonkin (Ed.)

More Small Astronomical Observatories
Patrick Moore (Ed.)

Observer's Guide to Stellar Evolution
Mike Inglis

How to Observe the Sun Safely
Lee Macdonald

The Practical Astronomer's Deep-Sky Companion
Jess K. Gilmour

Observing Comets
Nick James and Gerald North

Observing Variable Stars
Gerry A. Good

Visual Astronomy in the Suburbs
Antony Cooke

Astronomy of the Milky Way: The Observer's Guide to the Northern and Southern Milky Way (2 volumes)
Mike Inglis

The NexStar User's Guide
Michael W. Swanson

Observing Binary and Double Stars
Bob Argyle (Ed.)

Navigating the Night Sky
Guilherme de Almeida

The New Amateur Astronomer
Martin Mobberley

Care of Astronomical Telescope and Accessories
M. Barlow Pepin

Astronomy with a Home Computer
Neale Monks

Visual Astronomy Under Dark Skies
Antony Cooke

(continued after index)

Real Astronomy with Small Telescopes

Step-by-Step Activities for Discovery

Michael K. Gainer

With 90 Figures

Springer

Michael K. Gainer
P. O. Box 244
New Derry, PA 15671
USA
kizinski@aol.com

British Library Cataloguing in Publication Data
A catalogue record for this book is available from the British Library

Library of Congress Control Number: 2006927794

Patrick Moore's Practical Astronomy Series ISSN 1617-7185
ISBN-10:1-84628-478-3
ISBN-13:978-1-84628-478-6
e-ISBN-10:1-84628-508-9
e-ISBN-13:978-1-84628-508-0

Printed on acid-free paper.

©Springer-Verlag London Limited 2007
Apart from any fair dealing for the purposes of research or private study, or criticism or review, as permitted under the Copyright, Designs and Patents Act 1988, this publication may only be reproduced, stored or transmitted, in any form or by any means, with the prior permission in writing of the publishers, or in the case of reprographic reproduction in accordance with the terms of licences issued by the Copyright Licensing Agency. Enquiries concerning reproduction outside those terms should be sent to the publishers.

The use of registered names, trademarks, etc. in this publication does not imply, even in the absence of a specific statement, that such names are exempt from the relevant laws and regulations and therefore free for general use.

The publisher makes no representation, express or implied, with regard to the accuracy of the information contained in this book and cannot accept any legal responsibility or liability for any errors or omissions that may be made. Observing the Sun, along with a few other aspects of astronomy, can be dangerous. Neither the publisher nor the author accepts any legal responsibility or liability for personal loss or injury caused, or alleged to have been caused, by any information or recommendation contained in this book.

9 8 7 6 5 4 3 2 1

Springer Science+Business Media
springer.com

To my son Michael with love and gratitude

No unregarded star
Contracts its light
Into so small a character,
Removed far from humane sight,

But if we steadfast looke
We shall discerne
In it, as in some holy booke
How man may heavenly knowledge learne.

William Habington, 1634

Preface

The small refracting telescope with its simple direct design and construction and permanently aligned optics is the time tested standard for personal use. Easily portable, it can be taken out of the house and set up at a moment's notice. Or, it can be packed conveniently into a small car for traveling to a dark observing site. With moderate care and minimal maintenance, the telescope and mount can last for generations.

These instruments provide sharp, crisp high-contrast images of the Moon and planets. They are less susceptible to the effects of unstable air than larger ones, and because of their low maintenance and durability they are ideal for the parent educator or lifelong learner. Until recently, however, the high cost of small telescopes of sufficient quality for serious astronomical use has limited their accessibility.

During the past decade, innovations in optical design and manufacture have lowered cost and improved quality. Current developments in lens design make it possible to produce affordable refracting telescopes with shorter tube lengths and greater versatility. The once very costly 90-mm Maksutov–Cassegrain, with its optimum combination of effective optics and portability, now sells for a modest price.

I have often seen small instruments referred to as "grab and go" or "quick look," not worthy of consideration for serious observing. I maintain they can be much more. Despite their limits in types of observation, 80-mm refractors and 90-mm Maksutovs have sufficient aperture and magnifying power to provide a lifetime of observing pleasure. They are ideal for lunar occultation measurements, tracking the solar activity cycle, observing variable stars and measuring binary stars. Although all the activities described here are equally applicable to larger instruments, none of them require an aperture larger than 80 mm.

Through the use of common digital cameras, the limiting stellar magnitude and image scale of small telescopes can be increased dramatically, thus providing results one normally expects from larger instruments. Much of this book applies that principle to making interesting and useful astronomical observations. I have taken novel approaches to extending the use of small instruments for quantitative observations by applying unique methods for the analysis of digital camera photographs.

My intent is to demonstrate that useful and significant observations can be made with modest, relatively low cost equipment. Some of the activities described provide the necessary tools for making valuable contributions to various international astronomical data files. Others are of primarily educational value for either the self-learner or the science educator. This material should be of interest to both the beginning and the experienced observer. The emphasis is on what you can do with a small telescope rather than only on what you can see.

Acknowledgement

I am deeply indebted to Madalon Amenta for her time and skill in editing the manuscript.

Michael K. Gainer
June 2006

Contents

Preface ix

Computer Hardware and Software xv

Chapter One	The Celestial Sphere 1	
Chapter Two	The Measurement of Time.............................. 5	
	Solar Time 5	
	Sidereal Time 7	
	Dating Observations 8	
Chapter Three	The Equatorial Telescope Mount 9	
	Mount Stability... 9	
	The Polar Axis Drive..................................13	
	Setting Circles...13	
	Aligning an Equatorial Mount 15	
	Using Setting Circles..................................17	
	A Tabletop GEM 19	
	Go-to Mounts 20	
Chapter Four	Telescope Considerations................................21	
	Limitations Imposed by Aperture 21	
	Limitations Imposed by the Environment..............23	

	Limitations Unique to Refractors 23
	Short Focus Achromatic Refractors 24
	Limitations Unique to Maksutov Telescopes 25
	Newtonian Reflectors 25
	Eyepieces 26
	Focusing 26
	The Finder 27
	Recommended Accessories 27
	Which Small Telescope Should You Buy? 27
Chapter Five	**Astronomical Photography** 29
	Digital Photography 29
	Choice of Camera 31
	Mounting the Camera 31
	The Afocal Field of View 31
	Telephoto Conversion Lenses 32
	Processing the Print 33
	Film Photography 34
	Printing the Image 35
Chapter Six	**The Sun** 37
	Visual Observations of the Sun 39
	Classification of Sunspots 39
	Sunspot Number 40
	Visual Observations 41
	Digital Photography of the Sun 42
	Processing the Photographs 42
	The Stonyhurst Disk 43
	Measuring the Sun's Rotation 48
	Film Photography of the Sun 49
Chapter Seven	**The Moon** 51
	Visual Observations 51
	Digital Camera Photographs of the Moon 52
	Analyzing the Photographs 52
	Selenography 53
	Measurements on the Computer Monitor 56
	Measuring Lunar Libration 57
	Film Photography of the Moon 59
	Lunar Occultations 59
Chapter Eight	**The Planets** 61
	Sketching the Planets 61
	Filters 62
	Digital Photography of the Planets 62
	Plotting the Orbital Position of a Planet 64
	Ecliptic Coordinates 64
	The Phases of Venus 71

	Mars 71
	The Retrograde Motion of Mars 73
	Taking the Photographs 73
	Plotting the Results 74
	Visual Observations of Jupiter 74
	Digital Camera Observations of Jupiter 75
	Jupiter's Moons 76
	Roemer's Method for Measuring the Speed of Light 77
	Observations for Roemer's Method 77
	Saturn 78
Chapter Nine	**Comets and Asteroids** 79
	Comets 79
	Visual Observations 80
	Digital Photography 82
	Serendipitous Comet Discoveries 82
	Film Photography 82
	Asteroids 83
	Digital Photography 83
	Tracking an Asteroid 83
	Film Photography 89
Chapter Ten	**Visual Binary Stars** 91
	Digital Photography of Binary Stars 92
	Printing the Images 96
	Measuring the Separation of the Components 96
	Measuring Position Angle 98
	Film Photography 101
Chapter Eleven	**A Binary Star True Orbit Projector** 103
Chapter Twelve	**Visual Observations of Variable Stars** 109
	The Telescope 110
	Preparation for Observation 110
	Making the Observation 113
Chapter Thirteen	**Photography of Variable Stars** 117
	Processing the Image 118
	The Method of Measurement 118
	Making the Measurements 120
	Analyzing the Data 120
	Film Photography 122
Chapter Fourteen	**Star Clusters and Nebulae** 123
	Digital Photography of Star Clusters 125
Chapter Fifteen	**A Color-Magnitude Diagram for The Pleiades** 129
	Acquiring the Data 131
	The Analysis 132

Chapter Sixteen	The Design of an Objective Prism Spectrograph 133
	Getting the Spectrum 135
Chapter Seventeen	The Proper Motion of Barnard's Star 139
	Taking the Photographs 139

References and Additional Reading 143
    Star Atlases 143
    Telescopes and Accessories 143
    Astrophotography 143
    The Sun 144
    The Moon 144
    Binary Stars 144
    Variable Stars 144
    Star Clusters and Nebulae 144

Index 145

Computer Hardware and Software

An essential feature of this book is the application of computer processing and analysis with small telescopes to obtain data of scientific significance. If you do not already have an adequate computer and essential ancillary equipment, consider the following recommendations.

Hardware

The computer I have used for everything described here is equipped with an Intel Celeron 2.93-GHz processor, 512-MB RAM and an 80-GB hard drive. Slower PCs with less memory may have difficulty with some of the recommended software. The computer must have a sufficient number of USB ports to handle a scanner and digital camera downloads. Ports for various types of digital camera memory cards are more convenient than USB downloading. A scanner for scanning 35-mm slides and negatives is also essential.

Software

Although Guide 8.0 doesn't come with some of the bells and whistles and beautiful pictures of more expensive software packages, it is the best and least expensive software available for use with this book. It provides the ability to construct star charts with labeled magnitudes to the limit of any telescope and for any field of view. With it you can also label and identify variable stars and asteroids and obtain all the parameters necessary for observation of the Sun, Moon and Jovian satellites.

Microsoft Picture It! 7.0 or later editions is a low cost photo processor that has everything needed for the digital camera photographs associated with the activities in this book. With it a grid or scale can be superimposed over an image, faded to transparency and stretched and rotated through a calibrated angle. Any other software you might choose must have the same capabilities.

CHAPTER ONE

The Celestial Sphere

In order to make useful observations, it is necessary to establish a reference frame for making measurements. For all observations of the position of a star or planet to be related in time and space, certain directions and orientations must be established as standards. To establish such a standard coordinate system, the stars can be considered as fixed to a transparent celestial sphere that rotates from east to west about Earth's axis once every 24 hours. Figure 1.1 is an illustration of this imaginary sphere on which standard coordinates are indicated. The following definitions are referenced to this illustration.

The celestial poles are projections of Earth's poles onto the celestial sphere. The celestial equator is the projection of Earth's equator (e in Figure 1.1) onto the celestial sphere.

An observer at the point p on Earth sees a horizon indicated by the plane NWSE indicating the directions north, west, south and east. The observer's zenith, the point directly overhead, is p'. The local meridian is an imaginary line from the northern horizon, through the north celestial pole, through zenith, to the southern horizon.

As the Earth moves around the Sun at a rate of about one degree per day (360°/365.25 days), the Sun appears to move around the celestial sphere through the fixed stars at the same rate. This path of the sun is called the ecliptic. It is the projection of the plane of the Earth's orbit onto the celestial sphere.

The point at which the ecliptic crosses the equator, south to north going eastward, is the vernal equinox. The opposite point of intercept, 180° away, is the autumnal equinox. The most southern point on the ecliptic is the winter solstice;

2 Real Astronomy with Small Telescopes

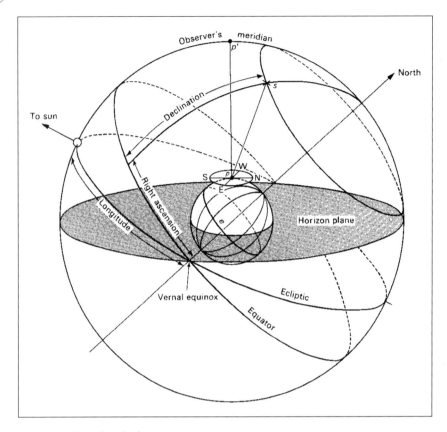

Figure 1.1. The celestial sphere.

the most northern, the summer solstice. Notice that the equinoxes and solstices are points on the celestial sphere not times of the year. Spring begins when the Sun crosses the vernal equinox not when the vernal equinox occurs. In Figure 1.1, the apparent Sun is indicated by a small circle at the winter solstice. The vernal equinox is on the eastern horizon. An arrow indicates direction to the actual Sun.

Due to the gravitational interaction between the Earth and the Moon, the Earth's axis precesses around a 23½° cone with a period of 25,800 years relative to the stars. As a result, the sidereal positions of the equinoxes change with the same period. For the purpose of establishing a coordinate system, however, we can consider them fixed.

Since the vernal equinox represents a fixed point on the celestial sphere, we use this point as the origin of a coordinate system to which we can refer the positions of stars. Suppose an observer at point p observes a star at point s on the celestial sphere. We can imagine a great circle running from north through the star to the celestial equator. We define the angle between this circle and the circle going through the vernal equinox as the right ascension (RA) of the star. For reasons explained in Chapter 2, right ascension is measured in hours, minutes and seconds (0 to 23 hours) eastward from the vernal equinox along the celestial equator.

We define the angular distance of the point s from the celestial equator as the declination of a star at that point. Coordinates of declination are parallel to the celestial equator. They are measured in degrees north or south. North declinations are $+0°$ to $90°$ and south $-0°$ to $-90°$. The coordinates of right ascension and declination, called equatorial coordinates, are equivalent to longitude and latitude on the Earth.

CHAPTER TWO

The Measurement of Time

Solar Time

In antiquity the time of day was measured by the direction of a shadow cast in sunlight. This resulted in the development of a wide variety of sophisticated and elegant sundials, which became the standard timekeepers. Sundials were also used as reference for other modes of time-keeping such as hourglasses.

Time kept by this method is called apparent solar time. The time between successive appearances of the Sun at the local meridian defines the apparent solar day. Because of Earth's elliptical orbit, the angular distance it covers per day varies. It moves more rapidly in winter when it approaches perihelion than in summer when it's near aphelion. As a consequence, the rate at which the apparent Sun moves eastward along the ecliptic, varies by the same amount. This causes the time between consecutive appearances of the Sun at the local meridian, the apparent solar day, to change as the year progresses. In ancient times this was not seen as a problem. But as more rigorously regulated civil activities and the expansion of intercontinental trade developed, a more consistent basis for time-keeping was needed.

Mean solar time was invented as a method for averaging out the inequalities of apparent solar time. Currently, it is defined in terms of a fictitious Sun that moves eastward along the celestial equator at a constant rate. The difference between mean solar time and apparent solar time for any particular date is called the equation of time.

For the convenience of civil activities, mean solar time is divided into global time zones that progress in increments of 1 hour, or approximately 15° westward, from the Greenwich meridian. For example, the Eastern Standard Time zone begins at 75° west longitude, making mean solar time for that zone 5 hours earlier than Greenwich.

The difference between time read on a sundial and a watch (mean solar time) in minutes is

$$\text{Apparent Solar time} = \text{Watch Time} + \text{Equation of Time}$$

The equation of time can be either negative or positive depending on the time of year.

Sundials can be constructed in a wide variety of forms and are an ideal medium for combining art and science. Devices can be constructed that compensate for the equation of time and longitude, and read mean solar time to an accuracy of one minute.

Horizontal sundials are simply the projection of the equatorial plane onto a horizontal plane as illustrated in Figure 2.1. The angle the gnomon makes with the horizontal plane equals local latitude. Vertical dials placed on a south-facing wall are constructed by the same principle. On those, the gnomon points downward opposite to the direction of the celestial pole. All forms of sundial must be aligned with the celestial pole in order to measure apparent solar time.

Another form of dial is illustrated in Figure 2.2. There the gnomon is the axis of a semicylinder that points in the direction of the celestial pole. The hour lines, 15° apart, are laid off parallel to the gnomon.

For those of you who would like to apply your esthetic sense to a practical artistic endeavor, more information can be found in the references at the end of the book.

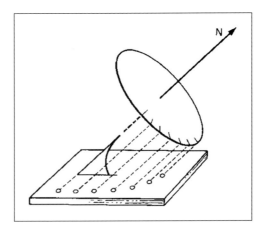

Figure 2.1. The horizontal sundial.

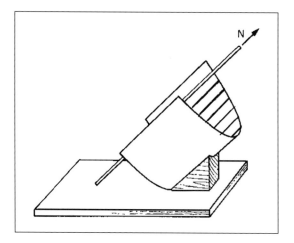

Figure 2.2. An equatorial sundial.

Sidereal Time

Apparent solar time and mean solar time are both related to consecutive appearances of the Sun or a representation of the Sun at the local meridian. Since the apparent Sun moves approximately one degree per day eastward relative to the stars, no form of solar time is useful in making measurements in reference to the stars.

The sidereal day is the time between consecutive transits of a particular star across the local meridian. Sidereal time is equal to the right ascension of a star at the meridian. Conversely, the right ascension of a star is the sidereal time at which it crosses the local meridian. This, in fact, is the way right ascensions are measured.

Because of the precession of the Earth's axis relative to the stars, however, the sidereal positions of the celestial poles, the celestial equator and the equinoxes change. In addition, each star has a proper motion relative to other stars. Consequently, the sidereal day is more rigorously defined as the time between the consecutive transits of the vernal equinox, thus tying it to a fixed reference frame. The position of the vernal equinox relative to the stars can be determined by measuring the time at which the Sun during its annual motion crosses the celestial equator from south to north.

The sidereal day is $3^m 55.91^s$ shorter than the mean solar day. Since civil activities are geared to solar time, sidereal time is only useful for astronomical observations.

Right ascension and declination are defined relative to the vernal equinox and the celestial equator. The equinox moves westward through the constellations of the zodiac in a 25800-year cycle. As a result, the cataloged values of equatorial coordinates for the stars are always referred to the position of the equinox for a particular epoch. For example, the positions for all stars in the current catalogs are referred to the equinox of 2000.0. The effects of precession and proper motion on the coordinates of a star are cataloged as annual variations in right ascension and declination. The exact current position is determined from the difference between the current time and the catalog epoch. Except for the nearest stars, the annual variations are fractions of an arc second.

Dating Observations

No astronomical observation is useful unless the time at which it was made is recorded. For observations at different geographical locations to be related, it is necessary for time measurement to be independent of the longitude or civil time of the observer.

Universal Time (UT), regulated by an international atomic clock, is mean solar time observed at the Greenwich meridian (GMT). Coordinated Universal Time is the time given by broadcast time signals.

Observations of phenomena that span long periods of time are dated by the Julian day, a period of numbered days devised in 1582 by Joseph Justus Scaliger and named after his father Julius Caesar Scaliger. Scaliger devised this dating scheme so that references to historic events would be independent of local calendars. Astronomers have adopted it for the same reasons. The Julian day for an observation can be obtained from the [Time Set] menu on Guide 8.0. A Julian day calendar can be downloaded from the American Association of Variable Star Observers (AAVSO) website.

CHAPTER THREE

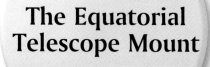

The Equatorial Telescope Mount

For any serious observing with small telescopes, an equatorial mount is essential. They are commonly used in one of two forms: the German type, referred to as the GEM, and the fork mount. The first is shown in Figure 3.1; the second in Figure 3.2. Each has two mutually perpendicular axes for positioning in right ascension and declination. In the most usable form, both axes are equipped with calibrated circles for positioning the telescope. The polar axis has a drive for tracking, and the declination drive has a manual slow-motion control. Suggestions and innovations for addressing some of the flaws and design problems associated with small telescope mounts follow.

Mount Stability

Since a shaking telescope is nearly impossible to bring into sharp focus, a rigid mounting is as important as are excellent optics. Slight vibrations produced by a gentle breeze or a hand touching the focusing knob will adversely affect observations. This is particularly troublesome with long focus models because of the greater torque produced by small forces at the eyepiece. Unfortunately, most imported 80-mm devices have inadequate mounts.

Figure 3.1 illustrates the standard equatorials that come with small telescopes. The one on the right usually goes by the generic name of EQ2; the one on the left, EQ3. Eighty-millimeter short tube refractors and 90-mm Maksutov–Cassegrains are frequently sold with the EQ1 mount. Lighter than the EQ2, it is inadequate

Figure 3.1. A comparison of German-type equatorial mounts.

for serious observing. Eighty-millimeter long focus refractors usually come with the EQ2. Although that mount has a solid base for a short focus refractor or a 90-mm Maksutov, it's not stable enough to dampen the vibrations induced by longer tube lengths.

Long focus refractors should be mounted, as the one shown in Figure 3.1, on the heavier EQ3. Orion sells the mount on the right as the AstroView. Other vendors sell an equivalent under different names. It's more stable than the EQ2 and has the added convenience of a built-in polar alignment telescope. Although a little heavier, it's still light enough to be easily portable. It costs about $70 USD more than the EQ2. The best solution is to buy a tube assembly and mount as separate items.

One of the advantages of the compact 90-mm Maksutov–Cassegrain is that it can be rigidly placed on an equally compact fork-type equatorial. The tube is mounted between the tines of a fork, as shown in Figure 3.2. A motor drive is built into the base of the mount. The base of the fork rotates about the polar axis, and an axis through the tines of the fork provides rotation in declination.

The tabletop legs provided with some fork mounts are convenient but they require a rigid table for satisfactory observing. Another disadvantage with this arrangement is that observing is awkward for some positions of the telescope.

Sufficiently accurate polar alignment for visual observations with tabletop mounts can be obtained by reference to a magnetic compass. But for photography, more precise alignment is necessary. A good method for achieving this is to wait for a sunny day and align the central leg of the mount with its shadow when the sun is at the local meridian. If you then mark the points where the

The Equatorial Telescope Mount

Figure 3.2. A fork-type equatorial mount.

feet of the legs touch the table, providing the table is not moved, you will have accurate polar alignment in azimuth automatically every time you use the telescope. Local mean time (watch time) at which the Sun transits your local meridian can be obtained from Guide 8.0, the RASC *Observer's Handbook* or any other astronomical ephemeris.

Alignment of the polar axis inclination is achieved by setting the declination circle to read your local latitude when the instrument is pointed to zenith. Place a level over the objective as shown in Figure 3.3 and adjust the inclination of the polar axis until the level is zeroed.

If you travel to a remote observing site, perhaps for a lunar occultation measurement, it is not likely that you will find a suitable table. For such occasions, a rigid portable tripod will be necessary. These are standard equipment with the Meade ETX. A few manufacturers are now selling the Maksutov as a spotting scope without a mount. One can then mount it on a separately purchased GEM.

Figure 3.3. Setting the inclination of the polar axis.

Figures 3.4 and 3.5 illustrate a portable, rock-solid tripod that I constructed for a Questar from parts picked up on Astromart. I fastened a Celestron C8 equatorial wedge to a standard aluminum tripod with a 3/8 inch bolt and wing nut for adjustment in azimuth. The arrows in Figure 3.4 point to two posts I made from copper sleeves and fastened to the wedge plate through existing holes. These serve to hold the mount base in place while being secured by the bar shown in Figure 3.5. The screw, used to fasten the bar to the base plate of the mount, carries a wing nut for tightening it to the wedge plate. This screw has a standard ¼ × 20 tripod thread and is long enough to permit passing the bar through the hole in the wedge plate without removing it from the base of the mount.

The Equatorial Telescope Mount

Figure 3.4. A Questar on a tripod made from spare parts.

The Polar Axis Drive

A drive is not needed for visual observations at low power. Neither is a drive needed for low magnification photography of the Sun and the Moon. But for photography of variable and binary stars, asteroids, star clusters and nebulae, a polar axis drive is essential.

For the digital photography described in this book, a drive that provides accurate tracking for a 30-second exposure is sufficient. I have found the drives available as optional accessories as EQ2 and EQ3 mounts serve the purpose nicely.

Setting Circles

The setting circles on many equatorial mounts currently available for small telescopes are not useful without some modifications. Sometimes you may need to bend the pointers with a pair of needle-nose pliers. Some mounts may require the addition of suitable indicators to make the scales readable. In Figures 3.6 and 3.7, otherwise unreadable right ascension and declination circles have been made more readable by adding an indicator cut from sheet brass, held in place by double-sided adhesive tape.

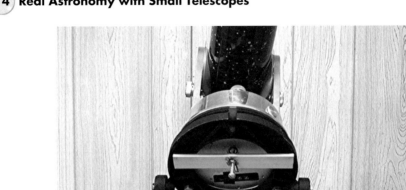

Figure 3.5. Rear view of the Questar mount.

Figure 3.6. An improved right ascension indicator.

The Equatorial Telescope Mount 15

Figure 3.7. A declination circle improvement.

Aligning an Equatorial Mount

The EQ3 mount shown in Figure 3.1 has a small telescope, which greatly simplifies the alignment of the mount with the north celestial pole, built into the polar axis. Most of these have a reticle in the eyepiece indicating the position of Polaris relative to the celestial pole. It's then a simple matter while sighting through the scope to adjust the mount in altitude and azimuth to proper alignment with Polaris. For mounts without this device, the procedure is as follows:

1. Set the telescope vertically above and parallel to the polar axis of the mount as shown in Figure 3.8. Lock the declination axis as shown in the illustration. Place a protractor level on the telescope, and adjust the altitude of the mount until the protractor reads your local latitude. Center Polaris in the field of the finder.
2. Rotate the telescope on the polar axis until it is parallel to the axis housing on either its east or west side as shown in Figure 3.9 and lock the right ascension axis. Rotate the telescope on the declination axis until Polaris is visible in the eyepiece.
3. Rotate the telescope mount in the horizontal plane (azimuth) and the telescope on the declination axis until Polaris is centered in an eyepiece. Lock the azimuth rotation axis.
4. Move the telescope back to the position in Figure 3.8 and adjust the inclination of the polar axis to bring Polaris into the center of the field of view.

Since Polaris is not exactly at the true celestial pole, this procedure will not point the polar axis precisely at the pole. However, for the observation periods and short photographic exposure times required for the activities described in this book, alignment with Polaris is sufficient.

Figure 3.8. Adjusting the inclination of the polar axis.

Figure 3.9. Adjusting the polar axis in azimuth.

The Equatorial Telescope Mount

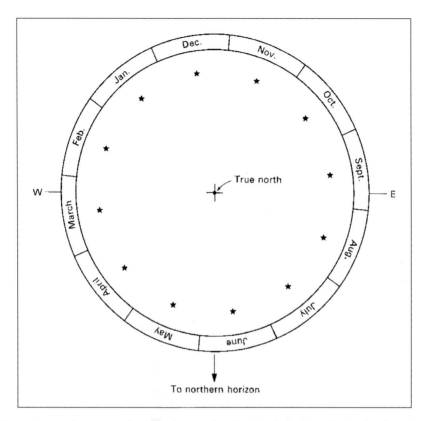

Figure 3.10. The position of Polaris relative to the celestial pole for the 15th day of each month at 0^h UT for 75° west longitude.

More precise alignment can be obtained by reference to Figure 3.10. It illustrates the position of Polaris relative to the celestial pole for the 15th day of each month at 0^h UT at 75° west longitude. For each degree west of 75°, Polaris should be rotated 1° clockwise on the chart. For each degree to the east, the rotation should be counterclockwise.

Using the method described in the previous section, rather than centering Polaris in the field, offset it by an amount indicated by the diagram. Polaris is located 0.73° from the pole. Knowing the apparent diameter of the field, estimate this distance from the center to Polaris. Lock the mount's altitude and azimuth axes. This method will permit sufficiently accurate tracking for 5-minute exposures with film photography using an 80-mm f/5 refractor.

Using Setting Circles

Before beginning an observing session, make a list of the right ascensions and declinations of the objects you intend to observe. Along with the object, list the position of a bright star near it. This will be a reference star for locating the target object.

To use the circles, first align the polar axis of the mount with Polaris by one of the methods already described. If it's the first time you're using the circles on the mount, check the calibration of the declination circle by the following procedure.

With the polar axis properly aligned, move the telescope to point to zenith and fix it in that position. You can do this by placing a level across the objective lens cell as shown in Figures 3.3 and 3.11. In that position the declination circle

Figure 3.11. Calibrating the declination circle on a GEM mount.

indicator should read the local latitude. If it does not, loosen the screw that secures the circle to the declination axis and move it to indicate the latitude. If the circle is not moveable, adjust the pointer.

To use the circle, center the reference star in the field of view and then move the telescope on the declination axis through the difference between the declinations of the reference star and the target.

For right ascension, place the reference star at the center of the field and move the right ascension circle to read its position. Then with the circle locked at that position move the telescope until the right ascension of the target object is indicated. Right ascension circles have two scales that read increasingly eastward. The top scale is for the northern hemisphere; the bottom, for the southern.

If the polar axis is accurately aligned, the setting circles on mounts similar to those shown in Figure 3.1 should be able to place an object within the field of view of a 25-mm eyepiece.

A Tabletop GEM

The convenience of the tabletop legs that come with some fork-mounted Maksutov–Cassegrain telescopes can be provided for small GEMs with what I call the "chopping block equatorial." The base of the mount is a heavy chopping block, measuring 11 × 15 × 2 inches, available at many kitchenware stores. The dense, hard wood is ideal. It dampens out vibrations and is too heavy to move if inadvertently hit by a wayward hand. Construction is simple: a matter of drilling holes in the tripod base of an EQ2 mount for attaching it to the board with bolts and wing nuts. A 90-mm Maksutov telescope is shown mounted on the finished equatorial in Figure 3.12.

Figure 3.12. A 90-mm Maksutov on a tabletop GEM mount.

Polar alignment can be achieved on any sunny day. Draw a perpendicular line from the horizontal edge of the block toward the center of the mount. When the Sun is at the local meridian align the north–south edge of the block with its shadow. At the same time adjust the mount on its azimuth axis so the shadow of the polar axis is perpendicular to the east–west direction of the block by referring to the previously drawn line. Once you have established alignment, you can mark the orientation of the block for nighttime reference.

With the GEM we avoid the awkward positions for observing circumpolar objects characteristic of fork-mounted equatorials. Simply slide the mount along the north–south line to the opposite side of the table and view from the south side. I have found that this mount when used on a solid table is more stable than conventional tripods for 90-mm Maksutov telescopes and 80-mm short tube refractors. Moreover, it provides a convenient table surface for reference charts, notebooks and accessories.

Go-to Mounts

I have heard mixed reviews from users about the effectiveness of go-to systems. It is probably just as easy and less frustrating to use alignment with Polaris and the setting circles on a GEM as it is to go through the two-star alignment process involved with setting up a go-to telescope.

For users of larger, permanently mounted telescopes high precision go-to systems are definite time savers. But, scaling down big astronomy is not always a wise thing. By looking for objects in the night sky you will see things that will arouse your curiosity. A little fuzzy patch of light or a small group of faint stars that you've never noticed before will beckon you to take a closer look. You will become a more astute observer.

Unless a Go-to mount is used as an equatorial, it cannot be used for long exposure photography. In the altitude-azimuth mode the computer will keep an object in the field of view but the field will rotate around it.

If you are a first-time telescope buyer you should consider whether some more worthwhile accessories, such as better eyepieces or a better digital camera, would be a wiser use of your money than a go-to mount.

CHAPTER FOUR

Telescope Considerations

Of the criteria I have chosen for small telescopes, the first is sufficient aperture. Others include optical and mechanical quality, durability, portability and versatility. The commonly available designs that best meet these criteria include 80-mm long and short focal length refractors, and 90-mm and 100-mm Maksutov–Cassegrains, hereafter simply referred to as Maksutovs. When the term "small telescope" is used in this book it will refer to these types and apertures. All photographs in this book were taken with 80-mm refractors and 90-mm Maksutovs.

Since optical configurations for these types of instrument are well known, the following discussion will center on the pros and cons of each as relevant to the activities presented here.

For all telescopes there are limitations in resolution, magnifying power and observable stellar magnitude imposed by the diffraction of light, atmospheric conditions and the physiology of the human eye. There have been many attempts to quantify those effects in order to calculate what to expect from a particular type of instrument of a given aperture. In reality, there is so much variation in the parameters for different types of optical system, geographic location and individual observers that they are, at best, approximate guidelines.

Limitations Imposed by Aperture

I live in the country, surrounded by farmland, 40 miles east of Pittsburgh, Pennsylvania, and 15 miles from any other major source of light pollution.

On the clearest moonless nights I can reliably detect stars as faint as 11.3 magnitudes with an 80-mm f/11 refractor. This is in accord with the relationship

$$m = 1.8 + 5 \log D$$

where D is the diameter of the objective in millimeters.

Although magnitude limits as faint as 12.1 have been quoted for 80-mm refractors, this is much too faint for the average observer. The value given above is more realistic.

Because of the obstruction of light by the secondary mirror, the effective aperture for a 90-mm Maksutov is 84 mm. Since there is additional light lost by two reflecting surfaces, the limiting magnitude for these telescopes is the same or slightly less than that of the 80-mm refractor.

The diffraction of light by a circular aperture imposes limitations on telescope resolving power and magnification. Diffraction produces an image of a star that consists of a bright central maximum surrounded by faint concentric rings. Faint stars at low magnification appear as points of light. For bright stars at high magnification the central maximum becomes perceptible as what is termed the "Airy disk." The diameter of the Airy disk decreases with increasing telescope aperture. If two components of a binary star are separated by a distance that is less than the diameter of the disk for a particular aperture, they cannot be resolved as individual stars. This criterion defines the resolving power of a telescope. It is given approximately in arc seconds by

$$R = 115/D$$

for D in millimeters. For an 80-mm refractor it is 1.4 arc seconds.

Telescopes are usually classified by the term "f/ratio." This is equal to the effective focal length of the objective lens divided by its aperture. An 80-mm refracting telescope having a focal length of 900 mm is designated as an 80-mm f/11.3 refractor.

While magnification of a telescope is equal to the effective focal length of the objective lens divided by the focal length of the eyepiece, for a Maksutov the effective focal length is the focal length of the combined mirror system.

Magnification can be changed by the use of eyepieces of differing focal lengths. But there is a limit to the magnification that can be achieved. At very high magnification the diffraction of light causes the image of a planet to loose contrast and sharpness. Consequently, the maximum magnification for a telescope depends on its aperture. In general 2× per millimeter of aperture is a good rule to apply for maximum useful magnification. On exceptional nights, when the air is extremely stable, this limit can be extended to 2.5× per millimeter.

The presence of the secondary mirror in Newtonian and Maksutov telescopes removes light from the Airy disk and distributes it to the secondary rings of the diffraction pattern. As a result the maximum useful magnification and image contrast are usually considered to be less than for a refractor. The effect is most noticeable for short focus Newtonians with relatively large secondary mirrors. It is negligible if the diameter of the secondary mirror, the diagonal minor axis for Newtonians, is 25% or less than 25% of the diameter of the primary mirror This occurs at about f/8 for Newtonians.

Design restrictions for Maksutov telescopes constrain the secondary mirror to a diameter of about 34% of the primary. However, their freedom from chromatic and other aberrations tends to offset the effect of diffraction by the secondary mirror. As a result, a 90-mm Maksutov can produce images comparable to those obtained with an 80-mm refractor.

Limitations Imposed by the Environment

Atmospheric conditions can seriously affect telescope performance. The stability of the atmosphere on a given night is described by the term "seeing," which is frequently assessed on a scale of I to IV. Since this scale is subjective, I prefer to use the qualitative terms "excellent," "good," "fair" and "poor."

In excellent seeing, the image of a planet distinctly shows fine details at better than 200× with an 80-mm refractor. In poor seeing, little more than the outline of a planet is visible and even that is poorly defined at modest magnification. Good and fair seeing conditions can be estimated by the highest magnification that will still yield a reasonably sharp image of a planet. The lower this is the poorer the seeing. Another qualifier of seeing is the lowest effective magnification that will resolve a closely separated binary star. On a night of excellent seeing, double stars with separations as close as 2.3 arc seconds can be resolved at 70× with an 80-mm refractor. With poor seeing they can't be resolved at any magnification.

First-time users of a reputable telescope shouldn't be discouraged if initially they don't see sharp images. The problem will more than likely be with the air rather than the optics. Try again when the air is more stable.

Limitations Unique to Refractors

The two primary imaging problems associated with refracting telescopes are spherical and chromatic aberration. Coma, astigmatism and field curvature may also afflict poorly made lenses.

The shorter the focal length, the greater these problems. Spherical aberration, coma and astigmatism can be controlled by properly figuring the two components of an achromatic lens. For a well-figured achromatic 80-mm lens with an f/ratio of f/11 or greater, chromatic aberration is negligible. Telescopes of this type produce high contrast, high resolution images. However, for f/ratios less than f/9 the color error becomes noticeable. And it cannot be corrected satisfactorily with a two-element achromatic lens. Although not bothersome or even noticeable at low magnifications, it can seriously deteriorate high magnification images in short focal length instruments.

There are two solutions currently being applied to this problem. The first and most effective is to add a third component to the objective. Proper figuring and adjustment of the three indices of refraction produces outstanding color correction for short focal lengths. With these apochromatic instruments, outstanding color correction can be achieved at low f/ratios. They are more portable and versatile than the long focus refractor but they are also much more expensive. Eighty-millimeter f/6 apochromatic tube assemblies cost ten times as much as a good 80-mm f/11 achromatic instrument.

The second and less expensive alternative is the use of extra low dispersion (ED) glass as one of the components of a two-element achromat. Most of the currently available 80-mm instruments of this type have focal ratios of around f/7.

Short Focus Achromatic Refractors

The performance of an 80-mm f/5 refractor can be greatly improved at high magnification by the use of a V-Block filter which effectively blocks blue and violet wavelengths while allowing 95% transmission at other wavelengths. With a V-Block filter combined with a 2.8× Klee Barlow lens, a good 80-mm f/5 refractor can be useful at a magnification as high as 150×. The photograph of the Moon in Figure 4.1 was taken with an 80-mm f/5 telescope using a V-Block filter with a 2.8× Barlow lens.

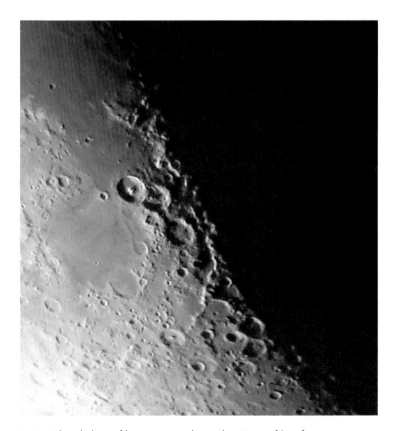

Figure 4.1. A digital photo of lunar craters taken with a 90-mm f/5 refractor.

For visual observations and digital photography of variable stars, comets, asteroids and star clusters, a short focal length refractor is more useful than an f/11. With a 25-mm Plossl eyepiece an 80-mm f/5 refractor will provide a field of view in excess of 3°. All of the pictures of variable stars, asteroids and star clusters in this book were taken with that type of telescope.

Limitations Unique to Maksutov Telescopes

The Maksutov–Cassegrain is a compound optical system with steep curvature elements. As a result, it takes much longer than a refractor to adjust to ambient temperatures. If taken from inside to outside in winter, a Maksutov will require at least an hour for acceptable images to be formed. An 80-mm refractor will require only 15 minutes.

The effect of poor seeing on star images is to broaden the Airy disk and increase the brightness of the secondary rings of the diffraction pattern. This is similar to the effect the presence of the secondary mirror has on the image. These two effects reinforce each other, making a Maksutov more sensitive to atmospheric conditions than a refractor.

Aperture-dependent performance of the 90-mm Maksutov is similar to that of an 80-mm refractor. If well made, the Maksutov is essentially free of image aberrations. Its greatest deficiency is lack of versatility. It is necessarily a long focal length instrument with a narrow field of view. The Maksutov excels for observations of the Moon, planets and binary stars but is more difficult to use for variable stars and other wide field applications. With good seeing and adjustment to ambient temperature, it is also a highly portable instrument. One must always, however, be aware of its limitations. In short, the Maksutov is an excellent choice for users living in moderately light-polluted areas whose interest is in solar, lunar, planetary and binary star observing.

Newtonian Reflectors

Newtonian reflectors are perfectly achromatic. With a parabolic primary mirror they are free of spherical aberration and their cost per millimeter of aperture is less than other optical systems. But they have limitations that restrict their versatility as effective small telescopes.

Although a parabolic mirror is free of spherical aberration, another type of image error, coma, restricts the useful field of view. This effect, which consists of elongated star images away from the center of the field, becomes greater with decreasing focal length. Another problem is that in order to produce a fully illuminated field the size of the secondary mirror has to increase as the f/ratio decreases, thus introducing diffraction effects detrimental to image contrast at high magnification.

This combined effect of coma and diffraction by the secondary restricts short focus Newtonian reflectors to low power observations. One can marginalize these detrimental effects if the instrument has an f/ratio of f/8 or higher. At such focal lengths a Newtonian can perform as well as other types of optical systems discussed here. The field of view of the 114-mm f/8 reflector is approximately the same as that of an 80-mm f/11 refractor and it has a fainter limiting magnitude.

The Newtonian has the same cooling-down problem as the Maksutov. In addition its open tube leaves the mirrors susceptible to deterioration from the environment and internal tube currents that can affect the image. The mirrors must be cleaned and realigned periodically. Consequently, the Newtonian requires substantially more maintenance than other types of telescope.

Perhaps the most serious problem with many of the 114-mm f/8 Newtonians currently on the market is that in order to reduce manufacturing costs they are made with spherical rather than parabolic mirrors. The spherical aberration they produce is not noticeable at low magnification but it severely affects contrast and image sharpness at high magnification. Since the distributors do not mention this in their advertising, the only way you can find out is to call and ask.

Eyepieces

Every bit as important to a telescope's performance as the objective lens is the eyepiece. It makes very little sense to have an excellent objective paired with an eyepiece that is inadequately corrected for image aberrations. It also makes little sense to use an eyepiece that costs as much and weighs half as much as the entire tube assembly.

The apparent field of an eyepiece is the field of view it would have if it were being used as a magnifying glass. The field of view of the telescope is equal to the apparent field of the eyepiece divided by the magnification that would be obtained with that eyepiece. For example, a telescope with a focal length of 900 mm will have a magnification of 36× when used with an eyepiece with a focal length of 25 mm. If the apparent field of the eyepiece is 52°, the field of view of the telescope will be equal to 52°/36 or 1.4°.

The choice of eyepiece for a particular observation depends on what is being observed. For the Moon, planets and binary stars use the highest effective magnification the telescope will take on a given night. For star clusters and nebulae, a low to medium magnification is best. For general use, the Plossl design with a 52° apparent field is a good choice. The 9-mm and 6-mm Orion Expanse eyepieces with a 66° apparent field and a relatively large rear lens are good for high magnification digital photography of binary stars, the Moon and planets. The Speers–Waler eyepieces provide a spectacular 82° apparent field but cost twice as much as the Orion Expanse eyepieces. They come in 18-mm, 14-mm, 10-mm and 7.5-mm focal lengths.

Focusing

The ability to precisely focus an image is another design factor that sometimes seems to elude telescope manufacturers. Good telescopes have a single, precise focal point where a sharp, high-contrast image is formed. Finding this point while observing can be frustrating, if not impossible, if the focusing device is poorly made. These difficulties often obfuscate otherwise superb optics. Don't buy a telescope with a cheaply made plastic focusing device. If you encounter vibrations induced by unsteady hands or mount instabilities, you can eliminate them by adding a battery powered focusing motor. These are available at modest cost from the sources listed at the end of the book.

The Finder

The finders that come with most commercial telescopes range from barely adequate to useless. Fortunately, most telescopes have a dovetail finder mount that allows easy replacement of a faulty finder. Any straight-through finder, regardless of quality, is difficult to use for objects near zenith with any type of telescope. It should be replaced with a good right angle finder with a minimum aperture of 30 mm. An 80-mm f/5 refractor doesn't need a finder; its 3.2° field with a 25-mm eyepiece is sufficient.

Recommended Accessories

Many of the following telescope accessories are useful but not all are essential for the observations described in the book. If you are an experienced observer wanting to get the most out of a small telescope you probably have them already and are familiar with their use. If you are a beginner, you can add them as your particular interests deem necessary.

As described in Chapter 9 an additional set of color filters for planetary observations provides higher contrast in certain aspects of their images. A 12.5-mm eyepiece with an illuminated crossline reticle is useful for some types of visual observations and for tracking long exposure film photography. A piggyback mounting bracket for a camera equipped with a telephoto lens is necessary for some digital and film photography. An electronic stopwatch is essential for lunar occultation measurements, timing eclipses and transits of Jupiter's moons and timing the transit of Jupiter's red spot. A red flashlight aids reading camera settings and recording data.

Which Small Telescope Should You Buy?

The answer to this question involves many variables. Here are a few: What is your environment, rural or urban? Are you an apartment dweller or do you have a backyard or garden? Can you easily travel to a less light-polluted area? Are you likely to take the telescope with you when you travel? Are you a science educator who wishes to have a versatile telescope for use with students? Are you a science educator who occasionally gives public presentations for which a portable telescope could be useful? Do you take a telescope with you, as I do, when visiting your grandchildren? Which activities described in this book are most likely to maintain your interest for an extended period of time? Which activities seem most useful to you if you are a science educator? What are the costs? What can you afford?

The appearance of excellent low cost 90-mm Maksutov telescopes on the market appears to have made the 80-mm f/11 refractor a disappearing breed. Orion sells a tube assembly designated as a guide scope for $120 USD. Meade offers one on an inadequate mount for around $200 USD. Both have excellent optical quality.

Several suppliers offer a good 80-mm f/5 refracting telescope tube assembly for under $200 USD. This is my favorite instrument for visual and photographic observations of variable stars.

Semi-apochromatic ED glass 80-mm refractors with focal ratios ranging from f/6 to f/9 are now appearing at prices under $500 USD. Apochromatic refractors of the same aperture are priced in excess of $1000 USD.

There is a considerable range of prices for a good 90-mm Maksutov instrument. An Orion 90-mm tube assembly mounted on an EQ2 with a drive will cost about $425 USD. The cost of the tube assembly alone is $220 USD. The Meade ETX 90 mm on a fork mount with tripod minus the go-to feature goes for $500. The minimum for a standard Questar is $4100. I have used each of these instruments and found them all to be of first-rate optical quality.

The Questar, an exquisite handmade masterpiece of heirloom quality, is unquestionably the jewel of small telescopes. On the other hand, if you are looking for a good functional telescope of that type, the low cost Orion Maksutov will serve just as well. All of the binary star photographs in this book were obtained with an Orion Apex 90 spotting scope mounted on an EQ3.

A good solution is to combine an 80-mm f/5 refractor and a 90-mm Maksutov with a rigid mount and drive. The cost of the two telescopes together is less than that of an 80-mm f/7 apochromatic refractor tube assembly, and the combination is more versatile. Both tube assemblies can fit interchangeably on the same mount and they use the same eyepieces. The system is highly portable and ideally suited for instructional purposes.

If you choose a single telescope and live in a moderately light-polluted area, I recommend a 90-mm or 100-mm Maksutov. Although its small field of view limits its versatility, it is a choice instrument for visual and photographic observation of the Sun, Moon, planets and binary stars.

The most versatile low cost choice for a single telescope is a good 80-mm f/5 refractor priced at less than $200 USD. If cost is not a major factor, an 80-mm f/5 or f/6 apochromatic instrument is a wise choice.

CHAPTER FIVE

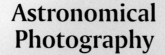

Astronomical Photography

Digital Photography

The common point-and-shoot digital camera has opened a wide range of possibilities for small telescope users. With it high magnification and high resolution photographs of the Sun, Moon, planets, asteroids and binary stars can be taken, processed and printed within minutes and then measured at leisure. Many high resolution 5-megapixel cameras currently available at modest prices are capable of low noise exposures for as long as 30 seconds at an ISO 400 equivalent.

In keeping with the basic criteria for small telescopes that is the focus of this book this discussion is restricted to fixed lens digital cameras. For astrophotography, I use the afocal method, which involves focusing the camera for infinity and photographing directly through the prefocused telescope eyepiece. With this technique, point sources such as star fields, open star clusters, binary stars and globular clusters can be photographed to a limiting magnitude dependent on the aperture of the telescope, the camera ISO and the length of a noise-free exposure. For Figure 5.1, the double cluster in Perseus, the camera was coupled to a 25-mm eyepiece on an 80-mm f/5 refractor. The exposure time was 30 seconds.

With digital photography, reciprocity failure – the decrease in film sensitivity as exposure time increases – does not occur. The diameter of a star image on a digital photograph is a linear function of the star's magnitude. This relationship is used for measuring the magnitudes of variable stars in Chapter 13.

On a clear moonless night, an 80-mm refractor can photograph stars as faint as 12 magnitudes with a 30-second exposure at ISO 400; a 15-second exposure will

Figure 5.1. An afocal 80-mm f/5 refractor photograph of the double cluster in Perseus. A 5.1-megapixel camera was combined with a 25-mm eyepiece.

reach 11.25 magnitudes. Within these limits, a number of long period variable stars can be observed through most of their light curves. Several of the brightest asteroids, those exceeding nine magnitudes at opposition, are easily within this range.

For diffuse nebulae and galaxies it's a different story. The object there is not a point source. The brightness of a point source at the focal point of the telescope's objective lens depends on its area. The brightness of an extended image at the focal plane decreases as the focal length increases because the light is spread over a larger area. As a result, the brightness of an extended object at the focal plane of a telescope depends directly on the aperture of the objective and inversely on its focal length, i.e., the f/ratio.

Although galaxies are aggregates of point sources, they are so far away that they appear as faint continuous sources in small telescopes. When a galaxy is described as being ninth magnitude, its brightness is that of a ninth magnitude star spread over the dimensions of the galaxy's image. The same is true of diffuse nebulae and comets. Thus, with a small telescope, all but the brightest of these objects require exposure times in excess of several minutes, well beyond the noise limit for fixed lens digital cameras. The equivalent of long exposures can be achieved by taking a number of short exposures and stacking them using appropriate computer software. This may require stacking 50 or more images to produce the desired result. Sources of information for such advanced imaging techniques are listed at the end of the book.

Short focus telescopes have an advantage over high f/ratio instruments for observing faint stars in a wide field. Unfortunately, they are also more efficient in intensifying background light. In moderately light-polluted areas an f/5 telescope may be less effective for faint extended objects than an f/8 or f/11. The advantage of an f/5 over an f/11 for digital photography is that it permits photography of a large field of view (2.5° to 3.0° degrees depending on the eyepiece) and the tracking of an object for a 30-second exposure without star trails. For satisfactory results, an f/10 instrument requires twice the tracking precision of an f/5.

Choice of Camera

To prevent undue stress and flexure when attached to the telescope eyepiece, the camera should be small and light. To be able to photograph all the detail that can be seen visually in the telescope it should have a minimum resolution of 3.1 megapixels. To provide a range of magnifications it should be equipped with a zoom lens and it must have either a remote shutter control or a self-timer so that vibration from pressing the shutter release by hand will not ruin the photo. Shutter speeds from 1/1000 second to 2.0 seconds are necessary for solar, lunar, planetary and binary star photography. If the camera is to be used for star clusters, variable stars and asteroids, an extended noise-reduced exposure capability of 15 to 30 seconds is also necessary. Photographs in this book were taken with a Sony W5, 5-megapixel camera, capable of 30-second exposures.

Mounting the Camera

If you ordinarily wear glasses but can clearly see objects at a distance without them, you don't need them to focus the telescope. If, on the other hand, you need your glasses for distance viewing, leave them on. This will assure that the telescope is focused for infinity and will provide a sharp photographic image with the afocal method.

Scopetronics manufactures adapters for almost any camera model for attaching a digital camera directly to the telescope eyepiece. An example is shown in Figure 5.2 for a Sony W5 camera.

The Afocal Field of View

The effective focal length of the combination of objective lens, camera lens and eyepiece is given by

$$F = F_o F_c / F_e$$

where F_o is the focal length of the objective, F_c the focal length of the camera lens and F_e the focal length of the eyepiece. For a 400-mm objective combined with a 7.9-mm camera lens and a 25-mm eyepiece, the effective focal length is 126.4 mm. The effect of the eyepiece in this example is to act as a focal

32 Real Astronomy with Small Telescopes

Figure 5.2. A digital camera with an afocal adapter on a 25-mm eyepiece.

reducer. The average fixed lens digital camera has a detector measuring approximately 7.2 mm × 5.3 mm. The width of the rectangular field of view for this combination is

$$D = (7.2/126.4)(180°/\pi) = 3.3°$$

For a 35-mm camera used at the prime focus of a 400-mm objective, the field is 5° wide. The focal length reduction for a digital camera is offset by the small size of the detector.

For a 90-mm Maksutov with a 1250-mm focal length combined with a 7.9-mm camera lens and a 9-mm eyepiece, the effective focal length becomes 3292 mm. The field is only 7.5 arc minutes. Here the eyepiece becomes an amplifying lens. Such long effective focal lengths are useful for photographing binary stars.

Telephoto Conversion Lenses

Rather than the afocal method through the telescope, a wider field can be realized with a digital camera by using a telephoto conversion lens. These lenses screw into the filter adapters of many digital camera models. The camera is then mounted piggyback on the telescope as shown in Figure 5.3.

Most of these lenses increase the focal length of the fixed camera lens by a factor of two. A typical digital camera lens has a focal length of around 24 mm when zoomed to 3×. A telephoto conversion lens will produce an effective focal length of 48 mm. With a 7.2 × 5.3 mm detector the resulting field width is 8.6°. This is equivalent to a 232-mm telephoto lens on a 35-mm film camera. The limiting photographic magnitude for a 30-second exposure with these conversion lenses

Astronomical Photography

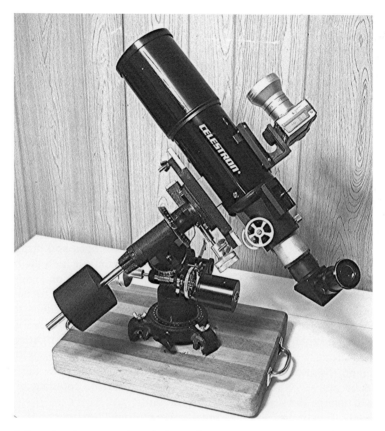

Figure 5.3. A digital camera with a telephoto adapter mounted on an 80-mm f/5 refractor.

is about eight. They are good for taking wide field photographs of constellations and producing a personal photographic star atlas.

Processing the Print

Always take several pictures of each object. Inevitably, some will be better than others. Occasionally a brief moment of outstanding seeing will yield a superb photo of a lunar surface feature when a dozen other photos of the same area will be mediocre. One photograph out of a dozen of a star cluster may be the only one devoid of star trails. The more shots you take the better your chances of getting a good measurable print.

Photographs of star clusters and asteroids may at first be disappointing. The computer thumbnail image may appear blank. If you increase the contrast, suddenly the stars will come out. Sharpen the image and even more stars may be resolved. Be careful to neither increase contrast nor sharpen excessively as this will bring out background noise. Inevitably wide field photographs will have some image distortion around the outer regions of the field that you can crop out. Cropping also magnifies regions of interest.

One can make several types of measurement more easily and more accurately on a print from a digital camera photo than with a micrometer eyepiece on a small or medium sized telescope. And on a cold cloudy night one can work in the comfort of a warm room. All one needs is a metric scale and a protractor.

One can also measure objects directly on a computer monitor, which can then be scanned and saved to a disk by using the overlay scales discussed in various chapters throughout the book. These scales include eight Stonyhurst grids for measuring sunspot latitudes and longitudes, a linear scale for measuring lunar surface features and a magnitude scale for measuring variable stars.

The procedures for using these overlays are described in the appropriate chapters. What is described here is based on the use of Microsoft Picture It! Premium 10. Other photo-processing software that will perform the same functions will serve as well.

To use an overlay keep the file of the processed print open and bring up the appropriate scale from the disk. Drag the scale to superimpose it on the image of the object to be measured and bring up the [Effects] menu. Click on [Transparent Fade] and then on [Even]. Fade the overlay to the desired degree of transparency.

Depending on the application, you may need to stretch or reduce the scale to fit the image. The scale can be rotated if necessary by any desired angle by clicking on the [Rotation] menu and then [Object] on the upper right corner of the menu. The angle of rotation can be read directly from the [Custom Rotation] indicator on the [Rotation] menu. When the process is complete, the scale or grid can be moved or rotated as desired with the cursor. Zooming in on the image will enlarge the scale by the same amount.

The application of the common digital camera to small telescopes has, in effect, transformed them into much larger instruments. They have their limitations but when their potential is fully utilized they are capable of much more than one might have expected.

Specific recommendations for procedures for various objects will be given throughout the text. Our purpose here is not to explain the basics of digital photography, but rather how to use it. Some excellent references on the details of digital imaging are listed at the end of the book.

Film Photography

The application of photographic emulsions for imaging and data acquisition along with stellar spectroscopy and electronic photometry are the technologies that made modern astrophysics possible. Although in recent years photographic emulsion has been replaced for most applications with digital technology there are still applications, such as wide field comet and asteroid searches and imaging nebulae and galaxies with small telescopes, for which film photography is more effective. The stellar spectra described in Chapter 16 cannot be done effectively with a digital camera. Unguided film exposures as long as 5 minutes can be made with a 400-mm lens at film speeds as high as 3200 ISO.

Astronomical Photography

As a result of the popularity of digital photography, excellent single lens reflex film cameras can be purchased at a fraction of their original cost. They are a good, economical choice for photographing nebulae, galaxies and wide field views of the Milky Way. For wide field applications film photography is best done with the camera mounted parallel to the telescope or at the prime focus of a short focal length refractor. A film camera can also be used with the afocal method described for digital cameras in the previous chapter. Adapters for attaching the camera directly to the eyepiece are available from Scopetronics.

The great disadvantage of film photography is that the images can't be seen until the film is processed and scanned to a computer. In that most commercial one-hour photo processors will no longer develop black-and-white film, the only alternative is to process it yourself, neither the most difficult nor the most time-consuming task. The major part of the work of film photography has always been printing.

Printing the Image

Contemporary printing is no longer the involved task it used to be. A darkroom with an enlarger, a variety of lenses and chemicals is no longer needed. Scanners with 35-mm slide and negative scanning capability provide high quality images for processing on the computer with next to no labor and expense. All of the overlays described here for digital cameras can be used with digitized film images.

The combination of film imaging and digital processing is a hybrid form of astronomical image acquisition. Astronomical film photography has existed for over a century. Consequently, there is a large volume of literature giving advice on such topics as choice of film, processing, lens, exposures and so forth. To go into such detail departs from the purpose of this book. This information can be found in the references.

CHAPTER SIX

The Sun

The photosphere, the apparent surface of the Sun, is the region of the Sun's atmosphere at which it becomes dense enough to be opaque at optical wavelengths. Its most fascinating aspect is the occurrence of sunspots, localized regions that appear black because they are cooler than their surroundings. Their origin is related to the continuously changing solar magnetic field.

The most curious phenomenon associated with sunspots is their periodic increase and decrease in number. This 11-year cycle has been occurring with consistent regularity over the 395-year history of solar observation, with a possible exception being a suspected 70-year minimum during the seventeenth and eighteenth centuries. Concurrent with the sunspot cycle is a 22-year solar magnetic polarity cycle.

With the rotation of the Sun, sunspots move from the observer's east to west, thus serving as markers for determining the rotation rate of the Sun as a function of solar latitude. At the beginning of a cycle, sunspots appear as far as 30° north or south of the solar equator. As the cycle progresses, their solar latitude decreases until at the end of the cycle they are generally located at or near the equator.

They also evolve and change their structure. Often, as a cycle maximum approaches, large spots associated in extended groups appear. These large groups are frequently the source of high-energy eruptions called solar flares. The protons and electrons ejected by flares enter the Earth's upper atmosphere and produce auroras. The most intense solar flares have been known to disrupt power grids and communication. Solar high-energy, in general, increases during sunspot maxima.

Another photospheric aspect is large irregularly shaped bright structures called faculae (Latin for "torch"). They appear brighter than the surrounding photosphere because they are a few hundred kilometers higher and about 300 K hotter. They are usually 15–20% larger than their sunspot regions. Faculae are most easily visible at the darkened limb of the Sun, near regions of recent sunspot activity or where sunspots are about to emerge. When faculae appear on the eastern limb of the sun, a group of spots will usually follow. Occasionally, polar faculae will appear farther north or south than the usual sunspot latitude, most frequently during the rise to a solar activity maximum.

Because sunspots are easily observed harbingers of increased solar activity, many attempts have been made to find correlations between sunspot cycles and terrestrial phenomena. Of particular interest is the possible relationship between solar activity and long- and short-term climate changes. Currently there are hints of correlations but no definitive cause-and-effect relationship points to a direct connection.

One example is the Maunder minimum. In 1890 E.W. Maunder, after a thorough examination of records, reported few sunspots documented between 1645 and 1715. Since the sun was assiduously observed during that period by some notable observers, the authenticity of the records could not be faulted by lack of attention. This 70-year period of low solar activity coincided with long, severe winters and short summers throughout Europe. Evidence indicates that the sunspot cycle possibly persisted during the Maunder minimum but the maxima consisted of only a few spots.

The intensity of the maximum, the duration of a cycle and its periodicity have varied over the history of observation. From 1761 to 1989 the time between maxima has been as short as 8 years, as long as 17, averaging 11.

In order to understand the nature of variations in solar activity, documentation during a minimum is as important as recording sunspot numbers during a maximum. Recording during a minimum may not be exciting but it is important and there can be surprises. Figure 6.1 taken on November 19, 2005, during the current sunspot minimum shows an anomalously large spot. There were only a few very small spots seen during the several weeks prior to and after the day that photograph was taken.

The long-term database that is currently available is insufficient to firmly establish definitive relationships between solar activity and terrestrial climate. The means of effectively detecting, monitoring and relating all of the relevant variables that effect climate change have not been available until the last quarter century. The technology for monitoring changes in solar activity, on the other hand, has been available to anyone who has had the inclination to use it for the past two centuries. It consists of a small telescope. This is one of the areas in which those interested can contribute important scientific data from backyard observations.

The minimum for the current solar cycle occurs in 2006. New solar activity cycles begin with the appearance of spots at high northern or southern heliographic latitudes. If you begin collecting data now you will be able to observe the onset of the next cycle and track it from minimum to minimum. Observing the next solar cycle can occupy your curiosity for at least 11 years.

The Sun 39

Figure 6.1. An unusually large sunspot observed during a solar minimum.

Visual Observations of the Sun

WARNING

Looking at the Sun with a telescope that is not properly equipped for solar observation can result in instant and permanent blindness. Always be sure an appropriate solar filter is securely in place before pointing the telescope toward the Sun. Never leave an unattended telescope set up in the daytime.

The traditional method of viewing the Sun has been to project its image onto a white surface attached by rods to the eyepiece end of the telescope. This has been supplanted by the use of special full aperture filters placed in front of the objective lens. These filters, made either of glass or mylar film, have metallic coatings that block ultra violet radiation and transmit only 0.0001% of the Sun's light. They are safe and provide a more detailed view than projection. Maksutovs must be used with a full aperture filter. Because of the internal reflections in a small closed system, the Sun's heat can damage the optical system if the telescope is used for an extended period with the eyepiece projection method.

Classification of Sunspots

Sunspots vary considerably in size and structure. Typically, they consist of a black center surrounded by a lighter penumbra, but they can appear as small single spots without a penumbra, as bipolar spots with a common penumbra

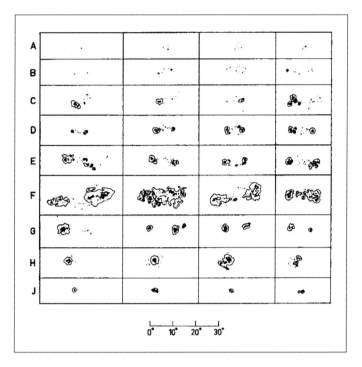

Figure 6.2. Sunspot types.

or as large bipolar groups subtending as much as 20° in solar longitude. In order to organize the spots according to their morphology, M. Waldmeir in 1938 devised what is now called the Zurich classification system. Representative appearances of each of the sunspot types in that system are shown in Figure 6.2.

Sunspot Number

The internationally standardized method of counting sunspots is to count the groups and the total number of spots, including those within each group. Individual spots, not a part of a defined group, are counted as a group.

If g is the number of groups of spots and N is the total number of spots, then, the relative sunspot number, R is given by

$$R = 10\,g + N$$

In Figure 6.3 two large groups are seen as well as some other groups that are labeled according to their type. These groups were the source of very intense solar flares.

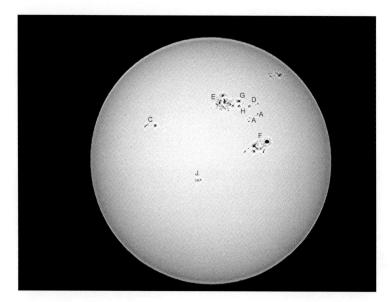

Figure 6.3. An 80-mm f/11 digital photo with sunspots labeled according to type.

Visual Observations

The best way to point a telescope so that the Sun will be in the field of view is to move it until the shadow cast by the tube forms a circle of minimum diameter. Using a low-power eyepiece, continue to move the telescope until the Sun is centered in the field of view.

The most important visual observation to make is to count the number of spots and the number and types of groups. A more detailed analysis of the positions and areas covered by each group and their changes in appearance can be done with digital photography, or one can also keep track by making sketches.

In order to define a reference frame for the positions of sunspots, the observer must establish east–west and north–south directions in the eyepiece. For equatorial fork-mounted Maksutovs, the north–south direction is pre-established because the telescope is constrained to move in that direction and the eyepiece is fixed. The east–west direction in the eyepiece can be determined by turning the drive off. The Sun will then drift across the field of view from east to west.

For telescopes on a GEM, the position of the right angle diagonal should be aligned with either the polar or declination axis. As with the Maksutovs, the east–west direction can then be determined by turning the drive off and observing the drift of the image across the eyepiece field.

Record the exact Universal Time at the beginning and end of your observation. Sketch, as accurately as possible, the location and nature of each group or single spot. Count the number of spots and the number of groups. By reference to Figure 6.2 determine the type of each group, and the number of each type. This should be done before photographing the Sun. The relative sunspot number thus recorded is the one that should be submitted for inclusion in the international database.

Observations of daily relative sunspot numbers can be checked by comparison with those in publications such as *Sky and Telescope* and *Astronomy Now*. After you feel confident in the reliability of your efforts, consider joining and submitting your observations to the Solar Observation Section of the AAVSO, ALPO, RASC or BAA. Your data will then become part of the international database for measuring solar activity.

Digital Photography of the Sun

Full aperture filters make digital photography of the Sun possible.

Caution: *Be sure the solar filter is firmly secured to the objective lens of the telescope. Use the same filter that was used for visual observing. Cover the finder objective lens.*

Make sure the frame of the camera view screen is accurately aligned with the north–south direction. To check this, move the telescope on the declination axis while rotating the camera until the Sun's limb moves parallel to its screen frame. Accurate alignment is essential for determination of the solar latitude and longitude of sunspots.

With the telescope precisely focused, lock the focusing mechanism and remove the eyepiece. Adjust the camera setting for the sharpest image and highest resolution. Set the shutter release for remote or self-timing. Focus the camera for infinity. For an 80-mm refractor or a 90-mm Maksutov, set the camera for manual control and an exposure time of 1/500 second at ISO 100. For larger telescopes you may have to use exposures as short as 1/1000 second.

Place the eyepiece with camera attached back in the telescope. A focused image of the Sun should now appear on the viewing screen. Because of flexure in the system the telescope may have to be moved slightly to center the image. By using the zoom control on the camera, the image size can be adjusted until it nearly fills the screen. Take at least 10 photographs to be sure to get some good ones.

Processing the Photographs

The polar axis of the Sun is inclined 7.15° relative to the plane of the earth's orbit. Consequently, its axis will be sequentially tilted toward the Earth, perpendicular to it, and tilted away from it by the same amount as the Earth moves around it. In addition, the Earth's equatorial plane is inclined 23.5° to its orbital plane. The result of the combination of these inclinations is described by two angles that change as the year progresses. The tilt of the solar axis toward or away from our line of sight is designated by the angle B_0. When the Sun's axis is pointing away from us B_0 has positive values. When it is toward us, B_0 is negative. The angle the Sun's polar axis makes with the north–south direction in the telescope is designated by P. This is given a positive value if the Sun's polar axis points east of the

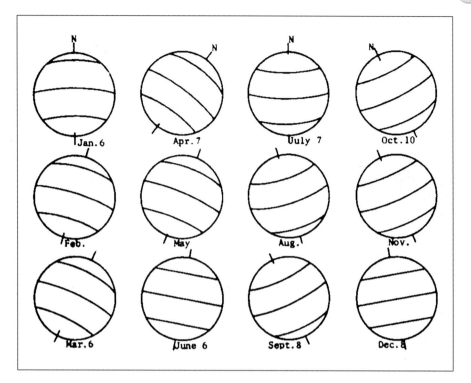

Figure 6.4. Monthly orientation of the Sun's axis.

north–south direction and negative if it points west. The orientation of the Sun's axis of rotation for different times of the year is shown in Figure 6.4. Use this chart as a reference for determining the orientation of the Sun in your telescope eyepiece.

The Stonyhurst Disk

Stonyhurst disks are grids for determining the heliographic longitude and latitude of a sunspot. They are plotted for both positive and negative values of B_o in steps of 1° from 0° to 7°. A complete set of these disks that may be scanned and saved to a CD is provided in Figures 6.5 through 6.13. They can also be obtained from ALPO and RASC websites from the *Oberver's Handbook* published by the RASC and various other astronomical almanacs.

The value of the central meridian on the disk is assigned the value of the heliographic longitude of the Sun's center for the Universal Time of the photograph. It is designated by L_o. Values of B_o, P and L_o for any UT can be obtained from Guide 8.0 and the RASC observers Handbook.

After the images have been downloaded, use the image-processing software to sharpen and improve the contrast to reveal as much detail as possible. Pick out the best one for analysis and download the Stonyhurst disk with the value of B_o for the date of observation. Make a print and save the image of the Sun before overlaying the Stonyhurst grid. Otherwise, it can't be recovered in its original form.

44 Real Astronomy with Small Telescopes

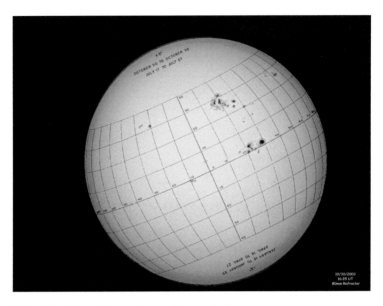

Figure 6.5. A Stonyhurst disk superimposed on a solar image.

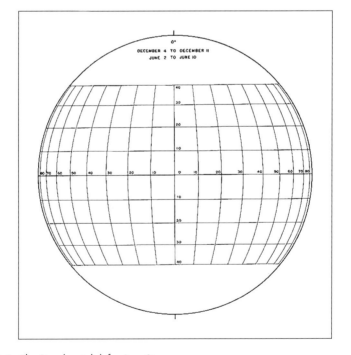

Figure 6.6. The Stonyhurst disk for $B_o = 0°$.

The Sun

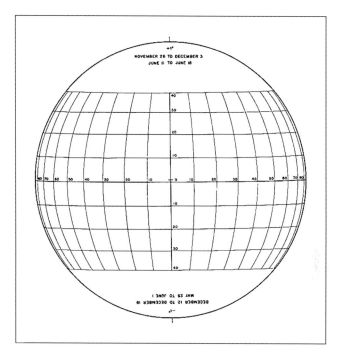

Figure 6.7. The Stonyhurst disk for $B_o = 1°$.

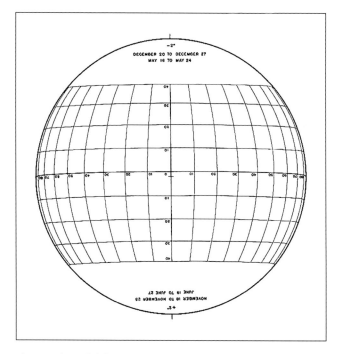

Figure 6.8. The Stonyhurst disk for $B_o = 2°$.

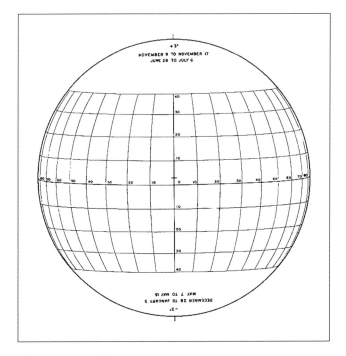

Figure 6.9. The Stonyhurst disk for $B_o = 3°$.

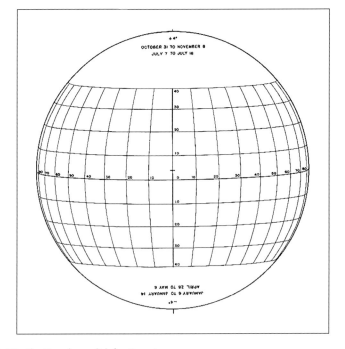

Figure 6.10. The Stonyhurst disk for $B_o = 4°$.

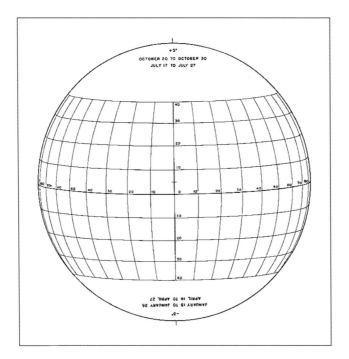

Figure 6.11. The Stonyhurst disk for $B_o = 5°$.

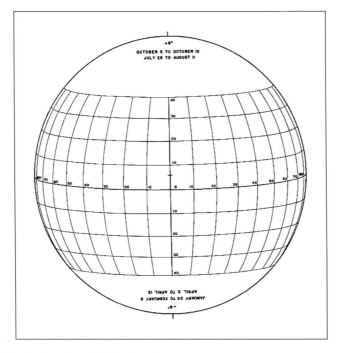

Figure 6.12. The Stonyhurst disk for $B_o = 6°$.

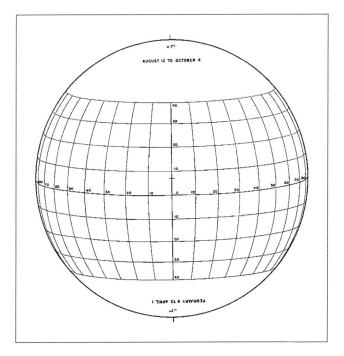

Figure 6.13. The Stonyhurst disk for $B_o = 7°$.

Bring up the Sun photo again. Drag the Stonyhurst disk to overlay the Sun's image. Using the software's [Effects] menu to work on the Stonyhurst image, transparent fade the image of the grid until the detail on the sun's image and the grid lines are both clearly visible. Then, using the [Even Fade] button, center the grid on the Sun's image. Now stretch the grid horizontally and vertically until it exactly coincides with the Sun's disk.

Go to the [Rotation] menu and click on [Object] in the upper right corner. Next, click on the [Custom Rotation] button. The grid can now be rotated through the angle P. For negative values of P rotate the grid clockwise. The angle indicator on the Picture IT! software screen increases for clockwise rotation. The indicated value will have to be subtracted from 360° for positive (counterclockwise) values of P. The result should be an image resembling Figure 6.5. The heliographic longitudes of the sunspots are given by $L_o + L$, and the latitudes are read directly. With care, one-degree accuracy can be obtained.

Save the processed image on a disk for a valuable, permanent photographic record.

Measuring the Sun's Rotation

The Sun does not rotate as a solid body. It takes 27 days to complete one rotation at its equator compared to an excess of 30 days at higher north or south solar latitudes. The rotation rate also depends on the solar cycle, which is slightly higher during a solar activity maximum.

The sun's differential rotation can be observed by timing the longitudinal motion of sunspots at different solar latitudes. This can easily be done by measurements with the Stonyhurst grid for successive images of the sun.

Film Photography of the Sun

Black-and-white or color film with an ISO of 100 or less should be used for photographing the Sun. The afocal method used with a digital camera is also the best method to use with a film camera. A 90-mm f/14 Maksutov telescope combined with a 50-mm camera lens and a 25-mm eyepiece will produce a 22-mm-diameter solar image. For a typical 80-mm f/11 refractor, using a 20-mm eyepiece, the image diameter will be 20-mm. These image sizes nearly fill the 24-mm width of a 35-mm film frame (24 mm × 35 mm). Adapters for attaching a 35-mm camera directly to the eyepiece can be obtained from Scopetronics. These fit the filter thread (usually 49 mm) of a typical 50-mm camera lens and can be fastened directly to the eyepiece.

Caution: *Be sure a suitable solar filter is firmly attached to the telescope objective before focusing and taking the picture. Use the same filter that is used for visual observations. Cover the finder objective lens.*

After you have processed the film scan it and save it to a disk. You can analyze the image with the same techniques used for digital photography.

CHAPTER SEVEN

The Moon

Ever since the Polish astronomer Hevelius produced the first detailed lunar atlas in the seventeenth century, the Moon has been examined, sketched and photographed with telescopes of all sizes. And now we've even explored it on foot. Because of this, interest in the Moon as an object for serious observing seems to have waned. Relatively few observers take the time to survey the vast richness in scenery and ever changing detail that can be seen with a small telescope, but for me there is no sight quite as spectacular as the entire visible surface of the Moon just past first quarter, magnified 100 times, filling the eyepiece of a telescope. Mountains, valleys, rills, faults, craters and plains are all displayed so clearly that you can pick and choose which one you want to visit and explore.

Visual Observations

For reference at the telescope, it is useful to have a laminated map of the Moon. An excellent inexpensive one with a ten-inch diameter lunar disk is published by *Sky Publishing Corp.*

To reduce the brightness of the image at low magnifications, a neutral density filter is recommended but a light green or light blue filter will do. At high magnification a filter is not needed. Start by examining the Moon with an eyepiece that will permit the entire image to be included in the field.

After having surveyed the Moon with low power, choose a site to explore at high magnification. Use the highest magnification at which detail is still sharp and well defined. Anything higher is empty magnification.

Every observation of the Moon presents a different vista. The Sun will be at a different position in the lunar sky, illuminating features differently than the last time you saw them. As a particular crater is observed throughout the night, shadows change their shapes and lengths. Objects and small craters previously barely visible may appear on the crater floor. This is particularly true for objects near the terminator.

Rather than going quickly from one object to another, pick out one or two features near the terminator and study the changes. Try sketching them, or at least, writing a description.

When observing the Moon, generally the first questions that come to mind about the craters are about size. How big are they? How deep? How high are those mountains? Of course the answers are in lunar atlases but it's more interesting to measure them yourself. We can determine their dimensions by measuring the lengths of their shadows when they are near the Moon's terminator on prints taken with a digital camera.

Digital Camera Photographs of the Moon

Focus the telescope with the appropriate eyepiece and lock the focuser. Remove the eyepiece, attach the camera to it and replace it, being careful to not disturb the focus. Experiment with combinations of eyepiece and camera zoom to produce the desired results.

Be sure to cancel the flash mode on the camera. Set the camera to focus for infinity, shutter priority, ISO 100 and remote or self-timing. Although the Moon appears to be solely shades of gray, do not set the camera for black-and-white photography. You may lose some detail but you can change the image to black-and-white when you process it in your computer. For photographs of the Moon at low magnification, depending on the phase, use shutter speeds from 1/80 to 1/50 of a second. Then take several photographs at each of several increasing magnifications. For the highest, the shutter speed may be as long as ½ second with an 80-mm refractor. Take high magnification photographs of overlapping sections along the terminator. To be sure, it takes some experimenting and not a little patience to become familiar with the procedure for getting excellent photographs, but is well worth the effort.

Analyzing the Photographs

After you have downloaded them, select the best low power and high magnification photographs. Sharpen them and adjust the brightness and contrast to represent the moon as you see it visually in the telescope. Make a full-page print of each of the best.

Use the print showing the entire moon to measure the diameters of a few of the larger craters. Then use those values to calibrate the high magnification prints from which shadow measurements will be made.

Here is the procedure: To measure the diameter of the moon, use a metric scale on the low power print and then measure the diameter of a few large dominant craters near the terminator; the larger the crater the smaller the error. The Moon's diameter is 3476 kilometers. If D is the diameter of the Moon in millimeters on the print and X, the diameter of the crater in millimeters, then the diameter of the crater in kilometers will be given by

$$d = 3476(X/D)$$

Figures 7.1 and 7.2 are examples of low and high magnification photographs taken with an 80-mm f/11 refractor using a 5-megapixel camera. Figure 7.1 was a 1/50 second exposure with a 20-mm eyepiece that has a 60° apparent field. For Figure 7.2, the camera was used at maximum zoom with a 9-mm, 66° apparent field eyepiece. The exposure time was 1/5 second.

Selenography

The height of a lunar mountain or crater wall can be found by measuring the length its shadow casts on the lunar surface or crater floor. In Figure 7.3, the distance L is the length of the shadow of an object, as seen from Earth, which has a height h.

$$h = L \sin \theta$$

The angle ϕ is the angle between the object casting the shadow and the terminator. The terminator is the dividing line between the sunlit and dark areas of the moon. Since ϕ is equal to θ we can write

$$h = L \sin \phi$$

Selenographic longitudes are measured from 0° to +90° if east, and 0° to −90° if west of the Moon's mean central meridian. Selenographic colongitude is the selenographic longitude of the morning terminator. From new Moon to first quarter, it goes from 270° to 360°. After the first quarter it continues to move eastward from 0° to 90°, 180° and back to 270° at the first quarter.

If the selenographic colongitude is designated by S then from new Moon to first quarter the longitude of the terminator is equal to 360° − S. From first quarter to full Moon the terminator's longitude is equal to the colongitude. From full Moon to last quarter, the terminator is at 180° − S. From the last quarter to the new Moon the terminator's longitude is S − 180°.

The angle ϕ is the difference between the longitude of the object and the longitude of the terminator. The selenographic colongitude for any time on any date can be found in the lunar ephemeris on Guide 8.0. It can also be found in the *Astronomical Almanac* or other astronomical ephemeredes. From this

Figure 7.1. A low magnification digital photo of the Moon taken with an 80-mm refractor.

Figure 7.2. A digital photo of the crater Copernicus taken with an 80-mm f/11 refractor.

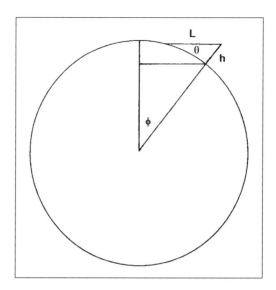

Figure 7.3. The geometric relationships for measuring lunar topography.

the longitude of the terminator can be determined. The longitude of the object casting the shadow can be found on a lunar map.

Use the high magnification photographs to measure vertical dimensions on the lunar surface. After bringing up the image, sharpen it and improve its brightness and contrast as needed. Use one of the craters already measured on low power to determine an image scale for the print. This will be equal to the crater diameter in kilometers divided by its diameter in millimeters. Measure the length of the shadow cast by the object of interest and convert it to kilometers using the scale. Determine the angle ϕ and calculate h.

The best measurements of lunar topography can be obtained near the Moon's equator and central meridian near the first and last quarter phases. There are a number of interesting surface features to measure. For starters, find and measure the heights of the Lunar Alps, the Lunar Apennines, the width of the Alpine Valley, the height of Mount Piton, the diameters of the craters Theophilus and Copernicus and the width of the Hyginus rill. It was suggested at one time that the craters Messier A and B located in Mare Fecunditatis near 50° east longitude, with rays streaming from them toward the west, were created by the entrance and exit of a single object. Measuring the distance between them should answer that question. Measure the width of the smallest clearly discernable feature on the print.

Measurements on the Computer Monitor

An alternative to measuring a print is to scan and download the linear scale that can be saved to a disk (Figure 7.4). First bring the photo to be measured, then bring up the scale and drag it to overlay the image. Fade the scale to transparency, move

Figure 7.4. A scale for measuring images on the computer monitor.

it to the location on the image you want to measure, then stretch or reduce it to the appropriate size. Zooming in on the object of interest will enlarge the scale by the same amount. Consequently, the calibration of the scale will not be changed.

Measure a crater on the lunar surface with a known diameter to calibrate the scale. Then you can measure any other shadow or object by moving and rotating the scale as desired. With this method one does not need to enlarge sections of the Moon by cropping the print. The scale can be moved around the Moon and zoomed in on any object of interest. In Figure 7.5 the scale is superimposed on the crater Copernicus.

Measuring Lunar Libration

Because the Moon travels in a slightly elliptical orbit, its orbital speed varies while its rotation rate remains constant, hence we can see slightly more than 50% of its surface in longitude. This is called libration in longitude. Libration

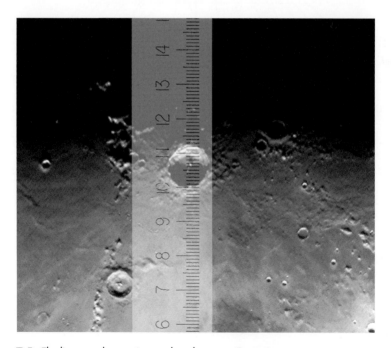

Figure 7.5. The linear scale superimposed on the crater Copernicus.

in latitude is caused by the inclination of the Moon by approximately 6.7° to the plane of its orbit. This combination of longitudinal and latitudinal librations permits us to see about 59% of the entire lunar surface.

Longitudes on the Moon are measured east or west from the mean center. They are positive from 0° to 90° if east; and negative, 0° to −90°, if west. Latitudes are measured from the lunar equator, positive if north; negative if south.

Lunar libration is measured by using the lunar coordinate grid in Figure 7.6. The procedure is the same as with the linear scale. Scan the grid, download it and then drag it to overlay the Moon's image. Fade it to transparency and then stretch it to fit horizontally and vertically. The lunar image will most likely be in a partial phase. In order to stretch the grid to a circle that fits the Moon's image, move it through four 90° rotations and refit it at each rotation. Adjust the rotation of the grid so that the terminator is parallel to the longitudinal grid lines.

The grid is a fixed reference frame against which east–west and north–south changes in the position of craters can be measured. Take photographs of the Moon as frequently as you can. It is best to photograph it after the first quarter or before the last. For each photograph measure the latitudes and longitudes of

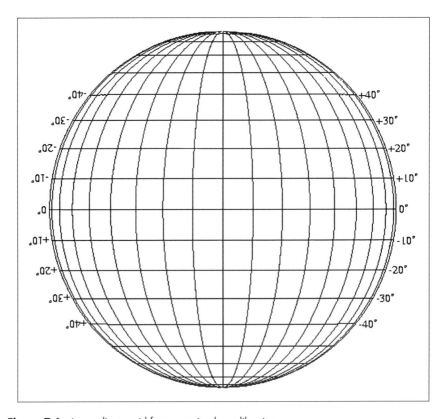

Figure 7.6. A coordinate grid for measuring lunar libration.

some select craters near the center and along its limb. Measurements made over a period of a few months will disclose the subtle dynamics of the Earth–Moon gravitational interaction.

Film Photography of the Moon

For film photography use the afocal method and the same procedure described for digital cameras. ISO 100 films with fine grain produce the sharpest images. With ISO 100 film, exposure times for the first quarter Moon should be on the order of 1/25 second with an 80-mm refractor. For high magnification the exposure time will need to be as long as ½ second. For each phase shoot a few images at several exposure times, keeping a record of the best exposure time for each phase as well sky conditions. This will save time and film during later attempts. The processed film can be scanned and analyzed on the computer using the procedures described earlier.

Lunar Occultations

As the Moon continues its 13° per day eastward motion relative to the stars, it often obscures background stars easily visible with small telescopes. Timing of these occultations provides a method for keeping track of the tidal evolution of the Earth–Moon system. The combination of tidal friction and conservation of angular momentum in the system causes the Moon to slowly recede and the Earth to gradually slow its rotation. In addition, the long-term effects produced by perturbations from other planets in the many body system can produce very small changes in both the Earth's and the Moon's orbits. Although these changes are trivial, a long-term record of them can verify some of the complexities of the theories for gravitational interactions.

The easiest occultations to observe with small telescopes are disappearances behind the Moon's dark limb. If the star is bright enough, observations of occultations by the bright limb of the full Moon are also possible with small telescopes. To time occultations an accurate time signal is necessary. These can be obtained from various short wave sources, for example WWV in the US at frequencies of 2.5, 5, 10, 15 and 20 MHz and CHU in Canada at frequencies of 3.330, 7.335 and 14.670 MHz. Clocks and watches that reset themselves with reference to a standard time signal every hour are other excellent sources. One will also need a good electronic stopwatch with large buttons, available at most sporting goods stores. For the measurement to be useful it is necessary to have accurate knowledge of your latitude, longitude and altitude, all to be found with a GPS receiver or geodetic survey maps of the area. Finally, the telescope should be equatorially mounted with a drive on the polar axis. The procedure is as follows.

Start the stopwatch at a time signal several minutes before the occultation is predicted to occur. Place the Moon in the center of a low power eyepiece and look for the star that is to be occulted. As the Moon approaches the star, increase the magnification so the star and the part of the Moon where the occulation is about to occur can be clearly seen. Observe carefully and attentively; the disappearance will be instantaneous. At the instant the star disappears, stop the watch, being careful not to clear it. Record the time lapsed from the start of the watch to the occultation. Add the elapsed time to the starting time to obtain the time of the occulatation. Record the time of the occultation and the latitude, longitude and altitude of the observing site. Report forms are available from the International Occultation Timing Association (IOTA).

Occultation predictions are published annually in *Sky and Telescope*, *The Observer's Handbook* published by the RASC and the IOTA.

Observing an occultation may require travel to a location where an event of interest is scheduled to occur. If you are willing and able to meet the demands of time and travel this can be a rewarding and useful activity.

CHAPTER EIGHT

The Planets

During the late nineteenth century and through much of the twentieth, detailed study of the surfaces of the planets was carried out with the great refracting telescopes at the world's major observatories. The space age has brought an end to this romantic era of astronomers spending endless nighttime hours sketching and photographing, hoping to catch a moment of steady seeing, in order to glimpse some evasive detail on the surface of Mars, Venus or Jupiter. We have walked on the Moon and our mechanical proxies have walked on Mars and landed on Venus. We have sent robotic emissaries to photograph and analyze the atmospheres of Jupiter, Saturn, Uranus, Neptune and their satellites with remarkable resolution and precision.

So what is left of interest for a person equipped with a small telescope other than an occasional look at the majesty of Jupiter with its retinue of satellites and the grandeur of Saturn and its rings? We can walk backwards through history and enjoy the re-enactment of discovery.

Sketching the Planets

Until the last half of the twentieth century, sketching the surface of a planet seen through a telescope was considered by many the best method for recording fine detail. The viewer could catch moments of excellent seeing in which minute detail that could not be detected photographically was clearly revealed. This method, by definition, is strongly dependent on the ability of observers to accurately and objectively draw what they see. Many, for example Antionadi, drew maps of Mars that accurately compare with recent spacecraft photographic maps of major albedo features. Percival Lowell, on the other hand, was much too imaginative.

It can be an interesting experience for you to do your own sketches. Since the image scale on photographs is too small to reveal the detail that can be seen with the eye, sketching can still be the best way to record an observation. Try to be objective and sketch what *you* see; not what you think you are expected to see. One of the most important benefits of this process is that it forces you to focus on detail that can turn you into a confident, critical observer rather than a mere looker.

Use the highest magnification that reveals a sharp high-contrast image. On some nights, higher magnification will be possible and more detail will be visible.

Filters

The use of filters that thread into standard eyepieces, available from most astronomical equipment suppliers, will enhance the appearance of details in the atmospheres of Venus, Jupiter and Saturn. Although cloud-covered Venus shows very little in the way of distinguishing features, a deep blue filter can sometimes bring out vague, slightly darker areas that reveal occasional irregularities along the terminator. For Jupiter, the boundary between belts and zones is sharpened by the use of a light blue filter; the appearance of festoons in the zones is enhanced with a yellow one. A light blue filter enhances the delicate contrast between Saturn's belts and zones.

Digital Photography of the Planets

Making measurements of the physical dimensions of a planet with small telescopes can be done more easily with digital photography than with a micrometer eyepiece. The micrometer requires experience, skill and a solidly stable telescope mount. Photographic prints can be measured in a warm room at your leisure. Measurable digital photographs of Jupiter, Saturn, Mars and Venus can be taken with an 80-mm refractor or 90-mm Maksutov using a camera with 3-megapixel or higher resolution. Although they won't reveal the rich detail of photos taken with larger instruments, they will provide enough information for some interesting activities. Figures 8.1 and 8.2 are representative photographs of Jupiter and Saturn taken with a 90-mm Maksutov telescope.

Exposure times will be from 1/25 second for Venus to 1/5 second for Saturn, depending on sky conditions and the brightness of the planet. Set the camera for an ISO of 100 and the highest resolution. Good results can be secured using a 6-mm wide field eyepiece and camera set at maximum optical zoom. With a 90-mm Maksutov, a 9-mm wide field eyepiece works well. Follow the instructions in Chapter 5 for setting up the camera. Take a number of photos at each of a few

The Planets 63

Figure 8.1. A digital photo of Jupiter taken with a 90-mm Maksutov telescope.

Figure 8.2. A digital photo of Saturn taken with a 90-mm Maksutov telescope.

shutter settings. Some will be better than others because of inevitable variations in seeing. Download all of the images to a computer and select the best for measurement. Store them on a disk for safekeeping and later use.

Plotting the Orbital Position of a Planet

An interesting activity is to determine the relationship between the position of a planet on the sky and its position in its orbit. This can then be extended to relating observations of the brightness and apparent diameter of a planet to its orbital position relative to the Earth.

Figures 8.3 through 8.6 are scale drawings of the orbits of each of the visible planets projected on the plane of the Earth's orbit. The scale on the outside circumference of each chart is labeled in degrees of heliocentric longitude, defined as the heliocentric angle, measured counterclockwise in the plane of the earth's orbit, from the vernal equinox to the planet. The direction labeled by the Greek letter Ω is the longitude of the ascending node. This is the angle between the direction of the vernal equinox and the point on a planet's orbit where it crosses from below the plane of the Earth's orbit to above it. The angular distance from the vernal equinox to the perihelion point on the orbit is designated by the angle ω. The orbit charts can be scanned to make multiple copies for later use.

Ecliptic Coordinates

The planet orbits all lie very nearly in the same plane as the Earth's orbit. Consequently it is sometimes advantageous to specify their positions with respect to the ecliptic rather than the celestial equator.

The geocentric ecliptic longitude is the eastward angle from the vernal equinox along the ecliptic, and is measured in degrees from 0° to 360°. For example, when the Sun is at the winter solstice, its right ascension is 18 hours and its ecliptic longitude is 270°.

Ecliptic latitude is the angle between the ecliptic and the observed object. Coordinates of ecliptic latitude run parallel to the ecliptic. They are measured in degrees, positive if north of the ecliptic; negative if south.

Figures 8.7 through 8.10 are charts on which both ecliptic and equatorial coordinates have been plotted. To find the geocentric ecliptic coordinates for a planet, plot its right ascension and declination on the appropriate chart and draw a line perpendicular to the ecliptic from the plotted point to the ecliptic. Where this line crosses the ecliptic will be the ecliptic longitude of the planet. Its ecliptic latitude can be read directly.

The heliocentric longitude of the Earth is 180° different from the geocentric longitude of the Sun. For example, when the Sun is seen at the vernal equinox its geocentric longitude is 0°. The Earth's heliocentric longitude at that time is 180°. To plot the position of the Earth in its orbit, first find the geocentric longitude of the Sun from Guide 8.0 or some other solar ephemeris. Plot this value on

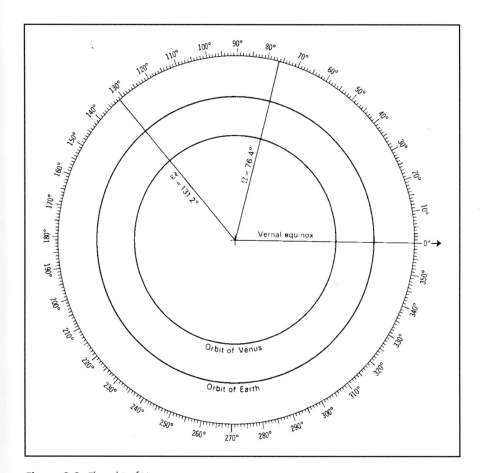

Figure 8.3. The orbit of Venus.

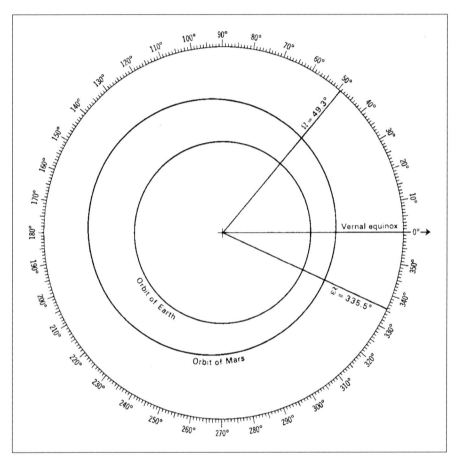

Figure 8.4. The orbit of Mars.

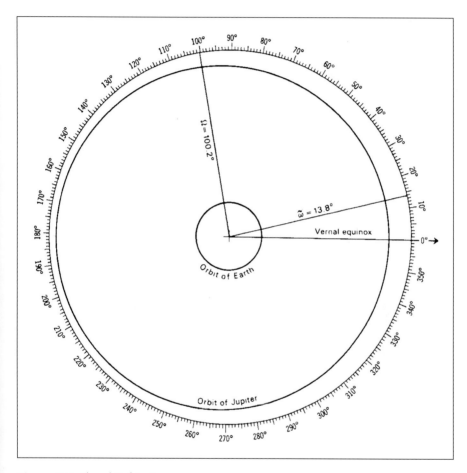

Figure 8.5. The orbit of Jupiter.

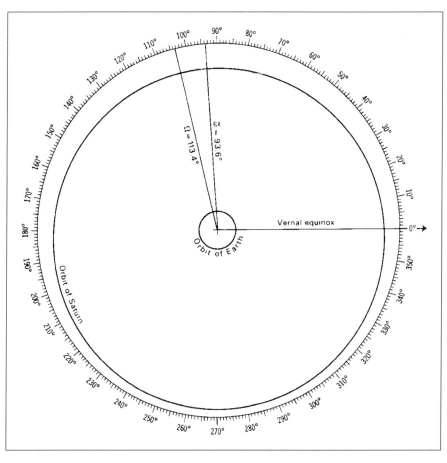

Figure 8.6. The orbit of Saturn.

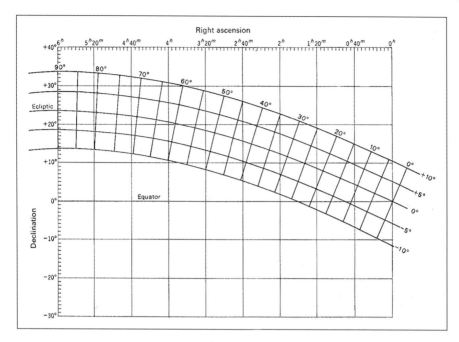

Figure 8.7. Ecliptic coordinates for RA = 0^h to 6^h.

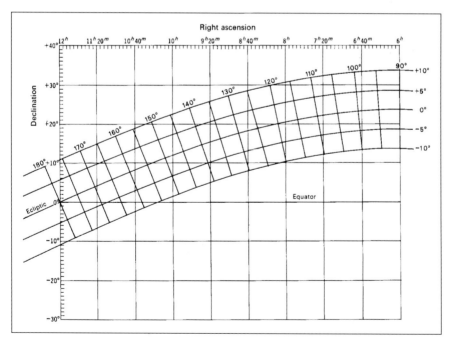

Figure 8.8. Ecliptic coordinates for RA = 6^h to 12^h.

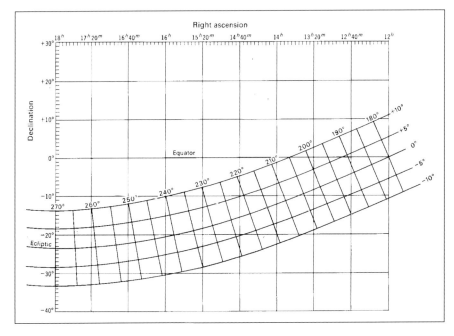

Figure 8.9. Ecliptic coordinates for RA = 12^h to 18^h.

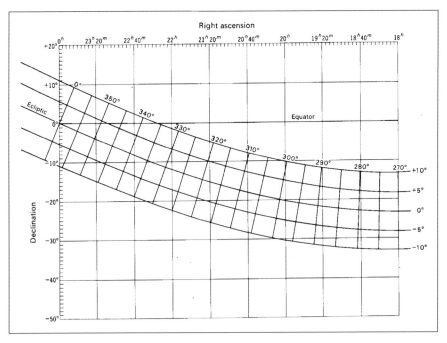

Figure 8.10. Ecliptic coordinates for RA = 18^h to 0^h.

the longitude scale on the circumference of the orbit chart for the planet that has been observed. Draw a line from this point through the Sun at the center of the chart to intersect with the Earth's orbit on the other side of the Sun. The intersection point will be the Earth's heliocentric position for the time and date of the observation.

To plot a planet's orbital position, you must first find its geocentric ecliptic coordinates by plotting its right ascension and declination on the appropriate ecliptic coordinate chart. Center a protractor on the position of the Earth with the 0°–180° direction parallel to the line joining the sun and the vernal equinox. Measure an angle, counterclockwise from the direction of the vernal equinox, equal to the geocentric ecliptic longitude of the planet. Draw a line from the Earth to that position. The point where the line intersects the orbit of the planet will be its approximate orbital position for that date. Figure 8.11 illustrates a finished orbital position plot for a date when the Sun's heliocentric longitude was 80° and Mars' geocentric longitude was 320°.

The Phases of Venus

One of Galileo's earliest discoveries was that Venus goes through phases similar to those of the Moon. The changes in the apparent diameter and phase of Venus can be observed by measuring a series of digital photographs of the planet, which can then be related to a plot of its orbital positions for those dates.

Take a set of pictures every two weeks, if possible, as Venus first appears as an evening star just after sunset, reaches its farthest point east of the Sun and then recedes westward before disappearing in evening twilight. If possible, take a similar series of pictures when the planet first appears as a morning star. Use the same camera and eyepiece combinations with exposure times on the order of 1/15 second at ISO 100 for all the photographs so that individual pictures taken on different nights can be compared. Record the time and date of each photograph and download them to a computer to save for later measurement. When you reproduce images, print each to the same scale. Science educators take note: You may find a set of these photographs useful in the classroom.

Measure the polar diameter of Venus on each print for each of the dates and note the phase. Using the procedure described earlier, plot the orbital position for each of the photos and relate the appearance of the planet to its orbital position for each point.

Mars

Mars can only be observed during oppositions that occur every two years. Even then, it is of little interest to the small telescope user unless the opposition occurs at or near its perihelion. These occur 2 years before, during and 2 years after Mars' perihelion passage. At those positions it can have an apparent diameter as large as 25 arc seconds. The planet's polar caps and most significant albedo features may be seen with small telescopes at that time.

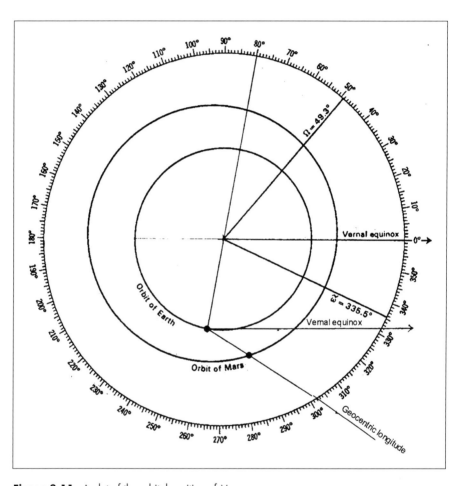

Figure 8.11. A plot of the orbital position of Mars.

The time between Martian perihelion oppositions is 17 years. The most recent close oppositions were in 2001, 2003 and 2005. The next close oppositions will be in July of 2018 and October of 2020.

At the next opposition on December 24, 2007, Mars will have an apparent diameter of 15.8 arc seconds. This is only about a third of the apparent diameter of Jupiter at opposition. It will continue to wane until March of 2012, when its apparent diameter will only be 13.9 arc seconds. For this reason I have excluded Mars as an object of interesting or useful visual observations with small telescopes.

The Retrograde Motion of Mars

The apparent motion of the planets among the stars is more complex than the casual observer may appreciate. When the superior planets, Mars, Jupiter and Saturn, pass through opposition they reverse the direction of their motion relative to the stars. They normally move slightly eastward every day. At opposition they change to a westward motion and then after a period of time, return to an eastward movement. In addition to this change in direction, the shape of the curve changes from opposition to opposition. This retrograde planetary motion was a perplexing problem to ancient astronomers as they attempted to develop a geometric model. It becomes particularly difficult if one assumes Earth is the center of the universe. In an attempt to solve the problem Ptolemy developed an elegant system of 89 cycles and epicycles, which the passage of time and improved observations found to be inaccurate. Eventually, the solution was found by Copernicus and refined by Kepler. They placed the Sun rather than the Earth at the center of the solar system.

The planet Mars, because it is the closest superior planet to Earth, demonstrates retrograde motion more obviously than the more distant Jupiter and Saturn. Its retrograde motion can be observed by taking a series of digital photos before, during and after opposition.

Taking the Photographs

Rather than using the afocal method, mount the camera parallel to the telescope as shown in Figure 5.5. This will provide a broad background of stars against which the motion of Mars can be traced. Carefully align the camera so that it is parallel to the length of the telescope. Center Mars in the telescope eyepiece and take exposures of 10, 15, 20, 24 and 30 seconds. After you download them decide on the best exposure time for future photos. Since recording faint stars is not necessary, 10- to 15-second exposures should suffice in acquiring enough reference stars to trace Mars' motion.

Proceed in the same manner over a period of time long enough to trace a substantial segment of Mars' path relative to the stars. For the 2007 opposition, begin in the middle of October and continue to mid-March 2008. If possible take a photo every two weeks. In this way you will record both ends of the retrograde loop. Be sure to note the right ascension and declination of Mars and the date and time of each photograph.

Plotting the Results

In order to make a visual record of the motion of Mars over the period of your observations make a transparency of the photo taken on the first date and plain paper prints of each of the others. Overlay the transparency on each of the other prints in turn and mark with permanent ink the position of Mars on the transparency. Notice that Mars will have moved eastward, then westward and then eastward again. Notice also that it will have changed its north–south position.

One can plot the orbital positions of Earth and Mars using the method described earlier in this chapter. After you have indicated those positions on the Mars orbit chart, draw lines connecting the positions of Mars and Earth for each observation. If you extend these lines to the end of the paper you will have the lines of sight to Mars for an observer on Earth. The distances to the stars relative to Mars are so great that any lines of sight to them from the different positions of the Earth will all be parallel lines.

Draw thin lines from each of the Earth's positions parallel to the line that connects Earth and Mars when they are nearest to opposition. These will be lines of sight to some star near the direction of Mars. By comparing the lines of sight to Mars to the lines of sight to the star, you will arrive at an explanation for the retrograde motion of Mars. The cause of the south to north motion, however, will not be so obvious.

Choose the appropriate ecliptic coordinate chart and plot the right ascension and declination of Mars for each of your observations. Note the positions of the ascending and descending nodes on the orbit chart, remembering that the orbit chart is drawn in the plane of the Earth's orbit. The ecliptic is this plane projected on the sky. The cause of the north–south motion of Mars should now become apparent.

Visual Observations of Jupiter

The surface of Jupiter has alternating dark and bright bands of clouds. The bright regions, designated as zones, are a higher cloud deck consisting of crystallized ammonia. The dark bands, called belts, lying deeper in the Jovian atmosphere consist of ammonium hydrosulfide. Although detailed observations of Jupiter's atmospheric activity are usually thought to require much larger instruments, there is enough visible in a small telescope to make careful observation and for sketching an interesting pursuit.

Observations of Jupiter are best made when the planet is near opposition and has its largest apparent diameter. The most prominent features visible in a small telescope are the north and south equatorial belts and the equatorial zone. Also commonly visible is the narrow north temperate belt. And with very good seeing other narrower, less distinct belts can be distinguished. The equatorial belts which may vary in width and contrast have borders of distinctly irregular appearance. Occasionally festoons, faint intrusions of the belts into the equatorial zone, are vivid enough to be seen. The southern belt has been known to disappear completely.

Because the appearance of the great red spot varies in intensity and hue, it requires excellent seeing to detect with a small telescope. Since it merges with the southern boundary of the south equatorial belt, it is usually visible as a faint oval protrusion into the zone south of it. It is more plainly visible when the south equatorial belt fades in intensity.

When the red spot is visible, a valuable observation is to time its transit across Jupiter's central meridian. If an eyepiece with a cross line reticle is not available, estimate when the spot is exactly at the center of the planet. Start a stopwatch at a standard time signal and stop it when you estimate that the red spot is at the Jovian meridian. Record the elapsed time and the time of the transit. Report your observation to the ALPO or equivalent organization. Predicted red spot transit times can be found in the RASC *Observer's Handbook*, *Sky and Telescope* and other astronomical publications.

Digital Camera Observations of Jupiter

When Jupiter is near opposition digital photographs of it can be useful for some interesting activities. One of the visual characteristics of Jupiter is the apparent difference between its polar and equatorial diameters. This seeming flattening of the poles could be an optical illusion due to the vivid horizontal banding of clouds or it could be a true physical difference between the polar and equatorial dimensions. We can determine which by photographing the planet and measuring its equatorial and polar diameters.

Proceed as described in the previous sections. Take several pictures, download them and save the best. Use a metric scale to measure the polar and equatorial diameters of the planet on your best print. They can also be measured on a computer monitor using the linear scale described in Chapter 7. The equatorial diameter of Jupiter is 142,800 km. You can use that value to establish a print scale km/mm and then calculate the polar diameter. These measurements can also be made, with the linear scale from Chapter 7, on a computer monitor. Figure 8.12 is an illustration of the scale overlaying the image of Jupiter.

In addition to sketching the appearance of the Jovian cloud belts, changes in their angular width and the latitudes of their borders can be measured by overlaying the planet's image with the coordinate grid from Figure 7.6. The procedure is the same as that used for measuring lunar libration.

Drag the grid to overlay Jupiter's image on the monitor. Then zoom in to sufficiently enlarge the image. Stretch the grid to fit the image and rotate it until the latitude lines on the grid are parallel to the cloud belts. Finally, measure the latitudes of the cloud belts and the angular widths of belts and zones. The image will be small but measurable.

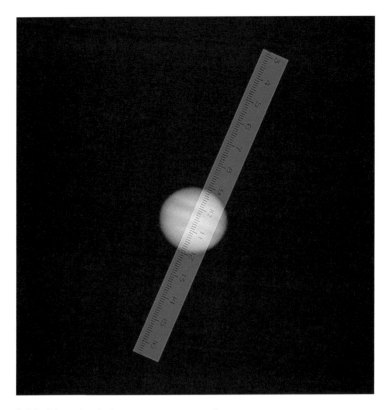

Figure 8.12. Measuring Jupiter on a computer monitor.

Jupiter's Moons

One of the most significant events in the history of science was Galileo's telescopic discovery that four smaller bodies orbit Jupiter. Arguably, all of modern astronomy, celestial mechanics and the way we think about the universe have their roots in that single event. We now know that Jupiter has 60 satellites. Even though our space probes have explored Io, Europa, Ganeymede and Callisto, they still fascinate us as we watch them being eclipsed by Jupiter or as we see their shadows transit its surface.

Predictions of occultations, eclipses and transits for the four Galilean satellites are on Guide 8.0, in the *Astronomical Almanac*, *The Observer's Handbook* of the RASC and in other popular astronomical publications. These events are not instantaneous. For example, it may take several seconds for a satellite to fade from view during an eclipse. Published prediction times are for the middle of the event.

Jupiter with its four brightest moons is ideal for studying gravitational interactions in a closed system. Timing satellite events provides the necessary data. By reporting your results to the ALPO you can contribute to the database. Contact them for more information.

Roemer's Method for Measuring the Speed of Light

The nearest approach of Jupiter to the Earth occurs when Jupiter moves through opposition. As the Earth continues to move in its orbit, the distance increases and Jupiter eventually disappears behind the Sun. At that time Jupiter and the Sun are said to be in conjunction. The planet is still visible, however, both a few weeks before and after. It is seen as a bright morning or evening star. At those times Jupiter is at its greatest visible distance from the Earth.

In 1675 Ole Roemer reported that the observed orbital period of Jupiter's satellite Io was shorter at opposition than when Jupiter was near conjunction with the Sun. He argued that the apparent differences in Io's orbital period indicated that light had a finite speed. The measured period was longer when Jupiter was farther away because it took longer for light to bring the information from the greater distance.

The additional distance light travels can be determined if the diameters of Earth's and Jupiter's orbits are known. The time for light to travel that distance is the difference between the orbital periods of Io at the two points.

Since Roemer's telescope was a refractor with a simple convex objective lens, his image of Jupiter was affected by chromatic and spherical aberrations. For timing he used a seventeenth-century clock of doubtful precision to more than a few seconds; and he depended on eye–hand coordination for his observations. Furthermore, the diameters of Earth's and Jupiter's orbits were not then accurately known. Is it any surprise that he came up with a value for the speed of light that is different from today's? His lasting distinction lies in the fact that he was the first to demonstrate that the speed of light is finite.

In theory, then, Roemer's method should be capable of providing a good approximation of the speed of light that is well within the capability of the modern small telescope. A modern small telescope is optically superior to the instrument he used. A contemporary handheld digital stopwatch yields accurate measurements limited only by the observer's eye–hand coordination, usually not greater than 0.2 seconds. Accurate distances to Jupiter can be found in readily available sources. Admittedly this is a difficult investigation to complete but I offer it as an activity that is a bit more challenging than others in the book.

Observations for Roemer's Method

The easiest events to observe are the beginnings of eclipses of a satellite by Jupiter. Predictions for eclipses of Io are given in the standard sources. Select from reference tables times for which the disappearing stage of an eclipse will be visible from your observing site. Try to find pairs of events separated by one Io orbital period, 1.76913 days. Then find two pairs separated by approximately three months from the first pair. One of the pairs should occur when Jupiter is first becoming prominent, rising in the eastern sky either late at night or early in the morning. The other should be obtained when the planet is near opposition. Find as many such pairs as you can so your opportunities for an observation will be less limited by uncontrollable events such as clouds. Possibly the most difficult and frustrating part of this activity is summoning the patience to wait for suitable observing conditions.

Begin your observation 30 minutes before the scheduled eclipse. After you have acquired Jupiter in the telescope field at low power, increase the magnification to the highest value at which it is sharply defined at high contrast. The eclipse will not be sudden. Io will begin to fade well before it approaches Jupiter's limb. At that time it's entering the penumbra of Jupiter's shadow.

Start a stopwatch on the signal provided by a UT time source. Observe Io as it approaches Jupiter and stop the watch at the instant you see it completely disappear. Record the elapsed time and determine the time for the beginning of the eclipse. From the time between two consecutive eclipses calculate the orbital period of Io. Repeat this procedure for pairs of eclipses when Jupiter is near opposition, and determine the difference between the orbital period observed near oppositions and conjunction.

Convert the UT of your observations to the Julian day recorded to an accuracy of 5 decimal points. The observed orbital period of Io will be the difference between the Julian day values of the successive pair of observations. The increased observed period of the orbit is the difference between the orbital periods determined for observations made months apart. Distances from the Earth to Jupiter for the observations can be obtained from Guide 8.0 and other astronomical ephemerides. Take the increased distance as the difference between the distances for the second of each pair of observations. The speed of light is equal to the increased distance divided by the difference between the observed orbital periods.

Using a simplified version of Roemer's method we cannot expect to obtain a value that comes to within 20 percent of the modern value for the speed of light. We can, however, observe as did Ole Roemer, that the speed if light is finite. We can also appreciate the difficulties intrinsic to the method and understand why historical references to values obtained with it cannot be compared to modern measurements with far more sophisticated methods.

Saturn

Saturn, even in a small telescope, is an astonishing sight. At 200× the major divisions in its rings are clearly visible and with good seeing the subtle banding of its clouds is easily apparent. The clouds vary somewhat in their visibility and so sketches we make of them, as with all planetary observations, should include a description of prevailing atmospheric conditions. This is particularly important for Saturn in order to separate the effects of poor seeing from any real changes in the appearance of its clouds.

An interesting measurement to make with digital photos of Saturn is the comparison of the diameter of the visible ring system with the diameter of the planet. This can be done either with prints or on the computer by using the linear scale described in Chapter 7.

CHAPTER NINE

Comets and Asteroids

Comets

New and previously discovered comets are frequently present in the night sky. Most are detectable only with large instruments but occasionally one may become bright enough for us to see with a small telescope. Then we can carefully observe changes in its appearance, brightness and position.

We can do this best with a wide field short focus refractor such as an 80-mm f/5 or f/6. An f/11 can be used with a low power wide field eyepiece for visual observations but photography with a long focus instrument requires precision tracking with an auxiliary guidescope. Unguided exposures of 30 seconds can be taken with short focus instruments if the mount is properly aligned with the celestial pole.

The apparent structure of a comet consists of two components, the coma and the tail. The coma, usually several thousand kilometers in diameter, consists of gas that has evaporated from the icy surface of a nonvisible nucleus which is only tens of kilometers across. The tail, which can exceed a million kilometers in length, frequently separates into two components: a dust tail that follows the comet along the direction of its orbit and, a gas tail that is displayed in a direction opposite the Sun. The tail precedes the comet as it moves away from the Sun; the direction is determined by solar radiation pressure.

Comets change their appearance as they come and go in their rendezvous with the Sun. The tail grows longer as it approaches, shorter as it departs. Sometimes a jet of gas will be ejected from the coma.

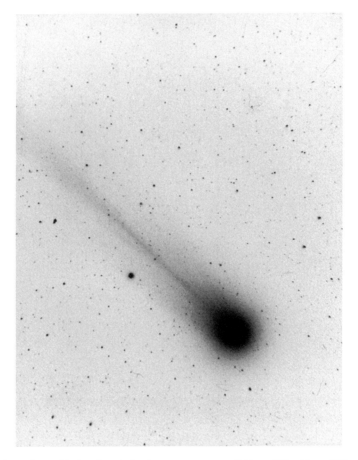

Figure 9.1. Comet Hyakutake: A 2-minute exposure on T-Max 400 with a 300-mm f/3.5 telephoto lens.

Appearance will also differ considerably from comet to comet. Figures 9.1 and 9.2 of comets Hyakutake and Hale–Bopp are examples. Note the double tail for Hale–Bopp and the differences in their comas. These photographs were taken with 2-minute exposures on T-Max 3200 film using a 300-mm f/3.5 telephoto lens before the widespread use of digital cameras.

Visual Observations

The positions of bright comets are published in all the popular astronomical periodicals. Using Guide 8.0 or equivalent software, make a star chart that has a field nearly equal to that of your eyepiece centered on the comet's position for the date of the observation. Label the magnitudes of the stars on the chart close to the comet, since they will be used to estimate the apparent magnitude. After aligning the telescope mount with the celestial pole, point the telescope to the brightest star near the position of the comet and set the right ascension and

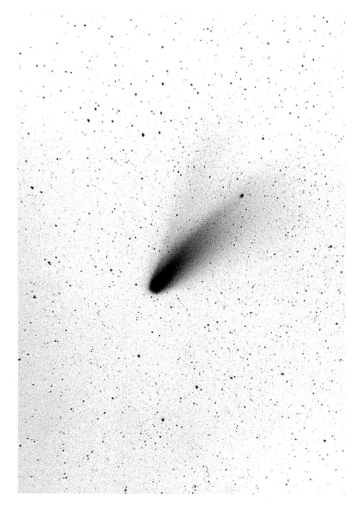

Figure 9.2. Comet Hale–Bopp: A 2-minute exposure on T-Max 400 with a 300-mm telephoto lens.

declination circles to the star's position. Then move the telescope to the position of the comet.

After you have identified stars in the eyepiece field that correspond to your star chart, pick out two or three you think will be good comparison stars. Defocus the telescope until the star images compare in size with the focused comet image. Using the stars that have a defocused image brightness most nearly equal to the focused comet image for comparison, estimate the magnitude of the comet.

Using an eyepiece that will provide a wide field and high contrast, sketch the appearance of the comet as frequently as observing conditions and time permit. In particular, look for changes in the appearance of the coma and tail. Note the distribution of the coma's brightness. Is it concentrated at the center or uniformly distributed? Record the time and date for each observation.

Digital Photography

When photographing, use an exposure time of 15 to 30 seconds at an ISO of 400 with the drive running. Use the eyepiece with the widest apparent field that is compatible with your camera mount. Any motion of the comet relative to background stars won't be detectable with such short exposures. Align the camera so that the image frame will be parallel to the east–west direction. Be sure to turn the flash off and use either a remote shutter control or self-timer. Note the exact time of each photo to the nearest second. Take several photographs. If possible take another set an hour or two later to track the comet's motion.

Prints will show more contrast and detail if they are converted to black-and-white negatives. Note date and time on the back of each and keep them for comparison of the comet images that will be taken at a later date. Be sure to archive the images on a disk.

The position of the comet can be estimated by bringing up the appropriate star chart on your computer and setting it to the same approximate field covered by the photograph. While comparing the computer chart with your photograph, place the cursor at the position at which the comet appears on the print. The coordinates of the position will then appear on the legend of the chart. Note this position on the back of the print and compare it with the predicted position.

If the comet has a visible tail, enhance the image contrast as much as possible without introducing excessive background noise. Measure the length of the tail and use a protractor to determine its orientation relative to the east–west direction. You can do this with a metric scale on a print. You can also make these measurements on the computer monitor using the scale described in Chapter 7.

Serendipitous Comet Discoveries

Since any wide field photograph you take may include some vaguely nonstellar object, it should be examined closely. Most (99.999%) of the time it will be something well known that you didn't realize was in that section of the sky. Most likely it will be a globular cluster, a galaxy, a small poorly resolved star cluster or a known comet. Check all references to establish its identity. If you can't identify it, take photos of the same region of the sky the next chance you get, to see if it's still there or if it has moved. If all efforts to establish its nature fail and you have detected motion you may have discovered a comet! You should report the details of your observation as soon as possible. The RASC *Observer's Handbook* gives information on how to report discoveries.

Film Photography

Film photography has the advantage of longer exposures with faster film. Use the camera at the prime focus of the telescope or use a telephoto lens and mount the camera parallel to the telescope as shown in Figure 5.6. If the telescope mount is accurately aligned with the celestial pole, unguided exposures as long as 5 minutes can be used with ISO 1600 film. Good choices for comet photography are 135-mm or 200-mm telephoto lenses.

Scan the negatives and download them to a computer. Analyze them with the methods described for digital camera images. Measurements are best made on a negative black-and-white print.

Asteroids

The four brightest asteroids can easily be photographed with an 80-mm f/5 refractor using a digital camera set at ISO 400 for a 15-second exposure. Asteroids as faint as 10 magnitudes can be easily recorded with 30-second exposures. Table 9.1 lists the four brightest with their maximum visual magnitudes.

The predicted positions for the brightest asteroids can be obtained from the *Observer's Handbook* by the RASC, the Astronomical Almanac or from Guide 8.0 and equivalent software.

Digital Photography

Guide 8.0 (or its equivalent) can be used to generate a chart, which shows the asteroid centered in a field equivalent to what you see in your telescope finder or in a wide field eyepiece. Use an eyepiece that provides a 2.5° to 3° field. By referring to the setting circles on the telescope mount, move the telescope to the asteroid's position. Compare the views in the finder and the eyepiece with your computer charts to center the asteroid in the field of view.

The photographic procedure is the same as that for comets described in the previous section. Use the longest exposure time for which your camera can be set at ISO 400. After you download the photograph you can identify the asteroid on a print by comparing it with the computer generated star chart. Figures 9.3 and 9.4 are examples of a photograph of the asteroid Vesta and the computer generated star chart for the date of the observation.

Tracking an Asteroid

You can track the positions of an asteroid on the sky over a period of a month or so using the following procedure.

Make a transparency of your first photo and record its time and date. Make plain paper negative prints of each succeeding observation. Overlay the transparency on each of the other prints in turn and mark the position of the asteroid on the transparency. Correlate these positions with the asteroid's orbital position by plotting its heliocentric position on the appropriate orbit chart (Figures 9.5 through 9.8). Determine their heliocentric longitudes by reference to Guide 8.0 or by the procedure outlined for planets in Chapter 8.

The computer generated position of the asteroid is its theoretically predicted position. In particular, gravitational perturbations produced by Jupiter have had to be taken into account for a precise prediction of its position. It is interesting to

Table 9.1.

Number and Name	Magnitude
1 Ceres	6.7
2 Pallas	6.7
3 Juno	7.4
3 Vesta	5.5

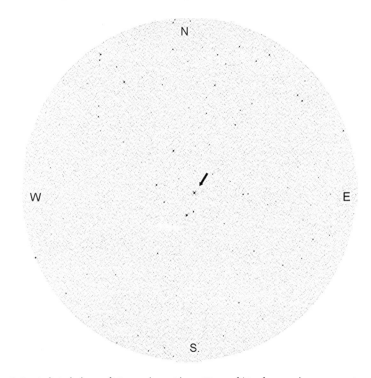

Figure 9.3. A digital photo of Vesta taken with an 80-mm f/5 refractor; the arrow points to Vesta.

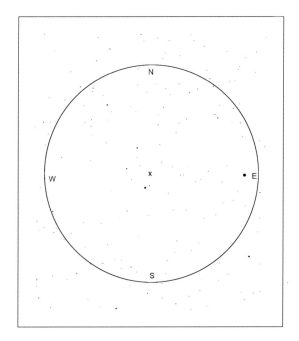

Figure 9.4. A computer-made chart for the position of Vesta. The "**x**" indicates its position.

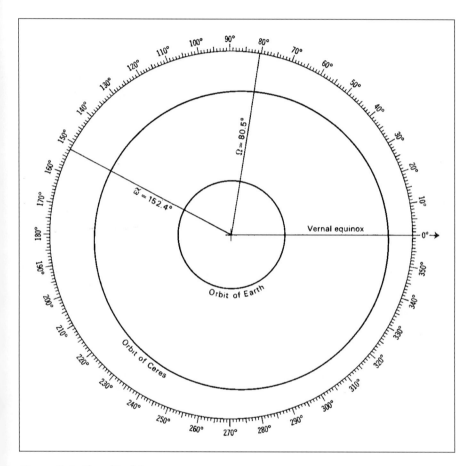

Figure 9.5. The orbit of Ceres.

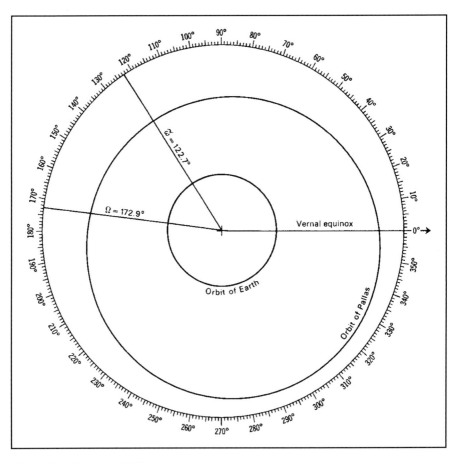

Figure 9.6. The orbit of Pallas.

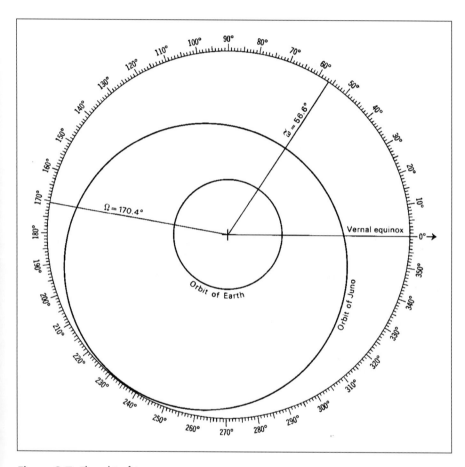

Figure 9.7. The orbit of Juno.

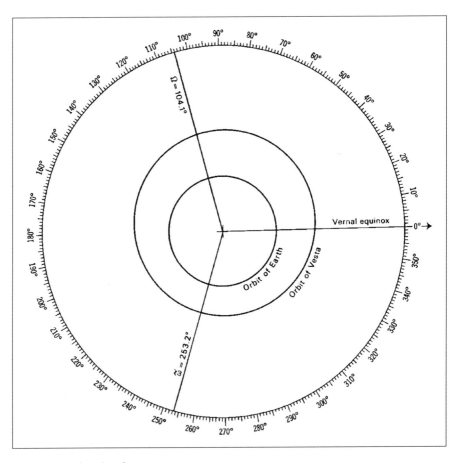

Figure 9.8. The orbit of Vesta.

compare this position with its position on your prints. Is the predicted position within the limits of your observational precision?

Film Photography

For film photography, use the same procedure and analysis that is described for comets in the previous section.

CHAPTER TEN

Visual Binary Stars

Observers in the last half of the eighteenth century noted the frequent occurrence of closely separated pairs of stars of nearly the same magnitude. Sir William Hershel verified the possibility that some of them might be gravitationally bound when in 1803 he produced evidence of orbital motion for more than fifty pairs. We now know that more than half of all stars are part of gravitationally bound binary or multiple star systems.

Two measurements, position angle and separation, are needed to determine whether or not a system is binary. These measurements, illustrated in Figure 10.1, are made with the brightest of the pair designated as the primary and the fainter as the secondary orbiting it. If the stars are of equal magnitude, the one with the highest right ascension is the primary. We measure the separation between the two in arc seconds, and the position angle in degrees from north in the eastward direction. When we have made a sufficient number of measurements, we can determine an apparent orbit of the secondary star.

From this apparent orbit we can apply projection geometry to determine the true relative orbit of the pair.

In developing theories of stellar structure, the determination of binary star orbits is second in importance only to measuring stellar parallaxes and absolute magnitudes. Applying the laws of gravity to these systems is the only way we can measure the mass of a star. Once their mass, color index and spectral class are determined, binary stars become standard masses to which all other stars of similar observational properties are compared.

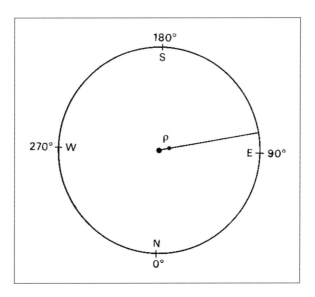

Figure 10.1. Binary star separation and position angle.

Many methods for making these measurements have been developed. The most frequently and traditionally used, the filar micrometer, is expensive, hard to find and not well suited to small telescopes. Also used are 12.5-mm illuminated reticle eyepieces that include a linear scale divided in increments of 0.1 mm and a protractor for position angle measurement.

For a 900-mm focal length refractor combined with a 3× Barlow lens, the 0.1 mm scale division of a micrometer eyepiece is equal to 7.6 arc seconds on the sky. As a result, only binary stars with separations significantly greater than 8 arc seconds can be measured accurately. Furthermore, such high magnification pushes the telescope to its useful magnification limit. Measurements of separations as small as 2.3 arc seconds, to an accuracy of 0.1 arc seconds, are possible through use of a common digital camera on a small telescope. Position angles can be determined with an accuracy of $1/2°$.

Digital Photography of Binary Stars

The requirements for digital cameras used in photographing binary stars are 3-megapixel or better resolution, remote or self-timer, shutter priority capability, 3× zoom and the ability to set the camera speed to ISO 400. Shutter speeds as long as 1 second should be available.

Figures 10.2 and 10.3 are photographs of the components of the double binary star system ε^1, ε^2 Lyrae. They were taken with a 90-mm Maksutov telescope using

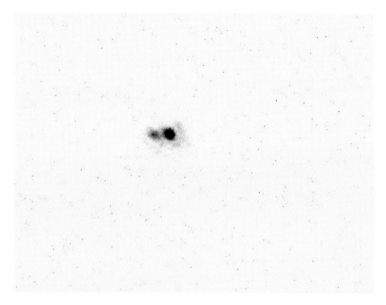

Figure 10.2. ε^1 Lyrae taken with a 90-mm Maksutov telescope; the separation is 2.5 arc seconds.

a 5.1-megapixel camera set at ISO 400. The shutter speed was 1/2 second. The separation between the components of ε^1 is 2.5 arc seconds. For ε^2, it is 2.4 arc seconds.

There are two prerequisites for obtaining good digital photographs. The first is a sturdy mount with a right ascension motor drive and a slow-motion control for declination, which does not have to be motorized. The second is good seeing. If the star images are not clearly resolved and steady you will be wasting your time.

Eyepieces with wide apparent fields greatly facilitate centering the star in the camera view screen. For all the binary star photographs in this book I have used Orion Expanse eyepieces with a 66° apparent field. With an 80-mm (focal length 900 mm) refractor, a 9-mm eyepiece combined with a 2× Barlow lens I get the best results. I have gotten similar results with a 90-mm Maksutov (focal length 1250 mm) using a 6-mm eyepiece without a Barlow lens. Taking good photographs of binary stars requires patience and attention to detail but once the procedure has been mastered it is not difficult. To photograph a binary star go through the following steps:

1. Before taking the photo, precisely focus the telescope on the binary star. Use the afocal method. Center the star in the eyepiece. With your hand apply downward pressure on the eyepiece approximately equal to the weight of the camera assembly and re-center the image. This will compensate for flexure in the system when the camera is attached. Once you have a sharp image, lock the focuser so the weight of the camera will not defocus the telescope.

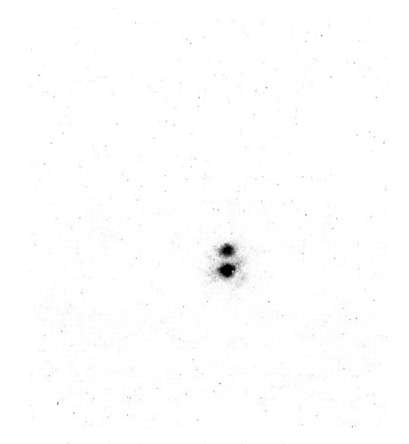

Figure 10.3. ε^2 Lyrae taken with a 90-mm Maksutov telescope; the separation is 2.4 arc seconds.

2. Set the camera for either manual operation or shutter priority. Set the focus for infinity, exposure time for 1 second and ISO at 400. Setting the exposure time for 1 second will make the star more visible on the viewing screen. You can set it for a shorter exposure when you take the picture. Depending on how your camera is equipped, use either a remote shutter switch or the self-timer. Be sure to turn off the flash or you'll be in for a big surprise.
3. Remove the eyepiece from the telescope and fasten it to the camera adapter. Place the eyepiece with camera attached back in the telescope. Align the camera so that one edge of the camera's viewing screen is parallel to the north–south direction and securely lock it. Do not zoom the lens until you are ready to start recording an image. The star, which will be a faint point of light, should be visible on the viewing screen.
4. With the star centered in the camera screen, carefully zoom the camera to its maximum (usually 3×). Do this slowly while keeping your eye on the screen to be sure the image is not lost as the field narrows while zooming. Do not use the digital zoom mode. You can gain greater magnification later when you process the image.

Visual Binary Stars 95

Table 10.1. Stars for Separation Calibration

Star	RA	Decl.	ρ (arc seconds)
Albireo	19^h31^m	$+33°52'$	34.6
Mizar	13^h24^m	$+54°56'$	14.6
κ Boo	14^h14^m	$+51°47'$	13.5
γ Vol	07^h09^m	$-70°30'$	14.1
ν Car	09^h47^m	$-65°04'$	5.0

Most of the stars listed in Tables 10.1 and 10.2 will require exposure times of 1/10 to 1 second, brighter ones such as Castor only 1/15 second. To be sure of getting a good image, use exposures of 1/15, 1/10, 1/5 and 1 second. Take several photos at each exposure time. After you have some experience with the process, the choice of exposure to use for particular stars and sky conditions will become more familiar.

The first picture you need to take is of a bright fixed pair such as Albireo, the head of Cygnus, or Mizar, the second star from the end of the handle in the Big Dipper. Mizar is actually a physical binary but its period, probably in excess of 1000 years, is so long that for our purposes it can be considered fixed. Their separations are given in Table 10.1. This photo will be used to calibrate your prints in arc seconds/millimeter. It must be taken with the same eyepiece and

Table 10.2. Binary Stars for Measurement

Star	RA	Decl.	m_1	m_2	θ (degrees)	ρ (arc seconds)
γ Ari	01^h54^m	$+19°18'$	4.5	4.6	1	7.5
λ Ori	05^h35^m	$+9°56'$	3.5	5.5	44	4.3
α Gem	07^h35^m	$+31°53'$	1.9	3.0	64	4.1
ζ Cnc	08^h12^m	$+17°39'$	5.1	6.2	73	5.0
φ² Cnc	08^h27^m	$+26°56'$	6.2	6.2	218	5.2
γ Leo	10^h20^m	$+19°51'$	2.4	3.0	123	4.5
α Cru	12^h27^m	$-63°06'$	1.3	1.6	114	3.9
Q Cen	13^h42^m	$-54°34'$	5.2	6.5	163	5.4
α Cen	14^h40^m	$-60°50'$	0.0	1.3	224	13.3
π Boo	14^h41^m	$+16°25'$	4.9	5.8	110	5.6
39 Boo	14^h50^m	$+48°43'$	6.3	6.7	47	2.9
δ Ser	15^h34^m	$+19°32'$	4.2	5.2	173	4.0
σ CrB	16^h15^m	$+33°52'$	5.6	6.5	238	7.1
ζ CrB	15^h39^m	$+36°38'$	5.0	5.9	306	6.4
ρ Her	17^h24^m	$+37°09'$	4.5	5.4	321	3.9
70 Oph	18^h06^m	$+02°30'$	4.2	6.2	139	5.1
ε¹ Lyrae	18^h44^m	$+39°40'$	5.0	6.1	350	2.5
ε² Lyrae	18^h44^m	$+39°40'$	5.3	5.4	82	2.4
γ Del	20^h47^m	$+16°07'$	4.4	5.0	26	9.2

camera combination you use for all your binary star photographs. If any of those parameters are changed, a new calibration star image will have to be established accordingly.

After you have photographed a calibration star, proceed in the same manner to photograph one or more of the stars listed in Table 10.2. After taking a series of photographs of a particular binary star, take a couple of 15-second exposures, with the telescope drive turned off at the beginning of the exposure. The resulting star trail will establish the east–west direction needed to measure the position angle for the system. Record the object, date, UT, exposure time and frame number for each picture.

Table 10.2 lists some stars that are possibilities for 80-mm refractors and 90-mm Maksutovs. None of these stars should require an exposure time longer than 1 second with the camera set at ISO 400.

Printing the Images

To measure the separation between components of a binary star, select your best photos of the calibration and program stars. Open also the image of the program star that was taken with the drive off. When first downloaded, the program star especially may be only faintly visible or, the primary may be visible and the secondary not. In some cases neither component will be visible. If you increase contrast, both components will be vividly displayed. Increase the contrast too much and you will obscure the separation. In order to make measurements, change the images to black-and-white negatives.

Measuring the Separation of the Components

To measure the separation between the components of the binary star use the following procedure:

1. Download the image of the binary star to your computer and select the best one. Convert the image to a black-and-white negative. Do the same for the calibration star.
2. Drag the image of the calibration star to overlay the binary. Transparent fade the calibration star image until it and the binary are both clearly visible and well defined.
3. Drag the image of the calibration star to be in close proximity to the binary and crop the final image to enlarge it for measurement and printing.
4. Using a small circle template as shown in Figure 10.4 draw a best-fit circle around each of the images and carefully mark the center of each. The final print should look like Figure 10.5. Using a transparent scale, measure the distance between the components of the calibration scale to the nearest ½ millimeter. A magnifying glass will be helpful.

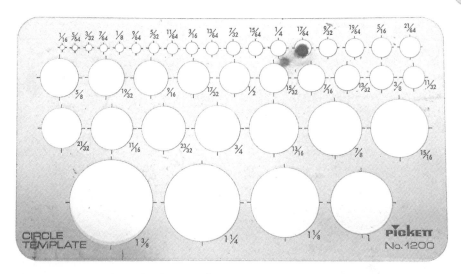

Figure 10.4. A circle template being used to measure ρ Her.

5. Use the above measurement and the known separation between the components of the calibration star from Table 10.1 to calculate the scale for the print in arc seconds/millimeter.
6. Measure the distance between the components of the binary and multiply that value by the print scale to obtain the separation ρ arc seconds.

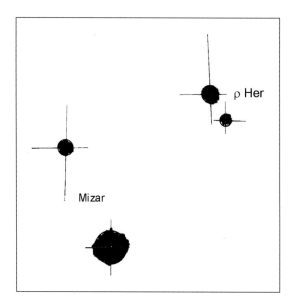

Figure 10.5. The calibration image of Mizar superimposed on the photo of ρ Her for separation measurement.

Measuring Position Angle

Convert the binary star image to a black-and-white negative and do the same for the star trail image taken with the drive off. Then open the binary star image and drag the star trail image to overlay it. Transparent fade the star trail so that it and the binary are clearly visible together on the screen. Then use the following procedure:

1. Drag the star trail so the primary of the binary is centered in it as in Figure 10.6. The star trail defines the east–west direction for determining the position

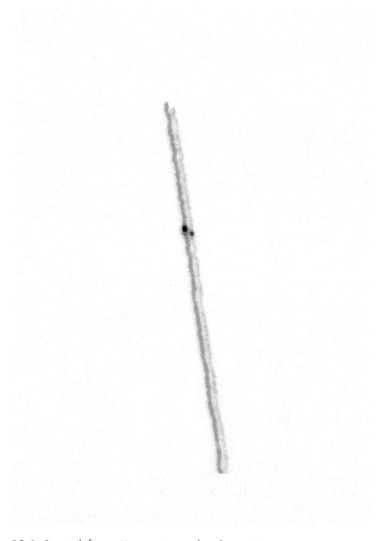

Figure 10.6. Star trails from ρ Her superimposed on the star's image.

angle, measured counterclockwise from north to east. As it drifts across the eyepiece field, a star moves from east to west so the beginning of the star trail is the 90° position angle. The opposite end is the 270° direction. Because a refractor produces a mirror image, it is necessary to flip the entire composite image horizontally before printing.

2. On the print, draw a fine line through the length of the star trail and extend it across the page. Draw another line connecting the centers of the binary star components and extend it across the page. When you're finished the print should look like Figure 10.7.

3. Use a circular protractor to measure the angle between the east–west line and the line through the centers of the binary star components with the primary star at the center. Using the convention for position angle illustrated in Figure 10.1 determine the position angle for the system.

Since drawing the lines for position angle is subject to errors in judgment, make several copies of the print and measure each. Use the statistical mode on a pocket calculator to determine the average value and standard deviation for the measurements of θ.

If you have drawn the lines and circles carefully you should be able to measure the position angle to an accuracy of better than one degree and the separation to 0.1 arc seconds.

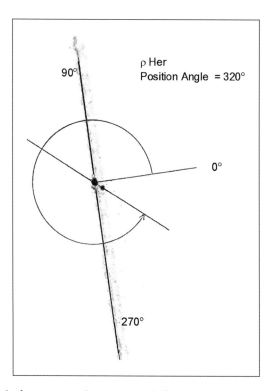

Figure 10.7. A plot for measuring the position angle for ρ Her.

Table 10.3. Measurements of ρ Her and ε^1, ε^2 Lyrae

Star	θ (degrees)	ρ (mm)	Scale (arc seconds/mm)	ρ (arc seconds)
ρ Her	321 322 319 320 319 321	3.7	1.09	4.03
ε^1 Lyrae	350 351 349 349 351 350 351	2.2	1.09	2.4
ε^2 Lyrae	81 80 80 82 80 78 82	2.1	1.09	2.3

Table 10.3 lists data from measurements of the binary stars ρ Her, ε^1 Lyrae and ε^2 Lyrae on images obtained with a 90-mm Maksutov with a focal length of 1250 mm. A 6-mm Orion Expanse eyepiece was used with a 5-megapixel camera zoomed to 3×. For ε^1, ε^2 the position angle of ε^2 relative to ε^1, rather than star trails, was used for reference.

Table 10.4 compares measurements of a few of the stars listed in Table 10.2, with values of θ and ρ predicted from the elements of their orbits for the observation dates. The predicted values are designated as θ_p and ρ_p. All of these observations were made with either an 80-mm refractor or a 90-mm Maksutov using the methods described here. The values are rounded off to the nearest degree, and 0.1″.

Although most of the stars listed in Table 10.2 will show next to no change in position angle or separation over the next several years, some will. You should not let the values in the table or those published elsewhere prejudice your results. The orbital elements for many of the binary stars suitable for observing with small telescopes are either poorly determined or not determined at all due to lack of sufficient data. Your observation may be at a much later date than the most recently published positions. There are three that will show measurable changes by 2010. Of these, 70 Oph will show the greatest because of its period of only 88 years. They are listed in Table 10.5.

An interesting star to observe over the next several years is γ Virginis. It was not included in Table 10.2 because the separation of its components is currently less than the resolving power of small telescopes. The secondary star moves in a highly eccentric orbit about the primary. It reached periastron in 2005 at a separation of 0.38 arc seconds. In 2010 the separation will be 1.46″, just within the resolution limits of an 80-mm refractor, and in 2015 it will reach 2.32″.

Table 10.4. A Comparison of Measurements with Published Values

Star	θ (degrees)	θ_p (degrees)	ρ (arc seconds)	ρ_p (arc seconds)	Telescope
70 Oph	138	137.6	5.0	5.0	90-mm Maksutov
α Gem	60	60.3	4.2	4.3	80-mm refractor
γ Leo	125	125.3	4.4	4.4	80-mm refractor
π Boo*	109	110	5.6	5.6	80-mm refractor
ρ Her	320	319	4.0	4.0	90-mm Maksutov
ε^1 Lyrae	350	349	2.4	2.5	90-mm Maksutov
ε^2 Lyrae	81	80	2.3	2.4	90-mm Maksutov

* The elements for the orbit of π Boo are not known. These values are from the *Washington Double Star Catalog* for the most recent observation (2002).

Table 10.5. Binary Stars with Noticeable Changes

Star	Year	Period (years)	θ (degrees)	ρ (arc seconds)
α Gem	2006	445	59.8	4.35
	2010		57.1	4.65
	2015		54.1	5.05
70 Oph	2006	88	136.8	5.06
	2008		134.6	5.42
	2015		131.6	5.72
γ Vir	2006	168.9	85.8	0.44
	2010		20	1.46
	2015		3.9	2.32

At that time the components will be easily distinguishable in small telescopes. They will continue to separate over the next couple of decades.

Film Photography

For film photography use the afocal method and proceed in the same way as with a digital camera. The single lens reflex type has the advantage of permitting visibility of the image of the star in the eyepiece as it is being photographed, thus making positioning of the image easier. With a 90-mm f/14 Maksutov, a 15-mm eyepiece with a 50-mm camera lens provides a good separation of the components. Exposure times with ISO film should be from 1/15 second to 1 second depending on the brightness of the star. After scanning the negative, analyze the prints using the same procedures used for digital camera prints.

CHAPTER ELEVEN

A Binary Star True Orbit Projector

By carefully measuring position angle and separation over a period of many years, we can determine the apparent orbit of the secondary star relative to the primary of a binary star system. It is the projection of the true orbit onto a plane perpendicular to the observer's line of sight. The terms used to describe the true orbit are called the elements of the orbit. They are defined as follows.

a = semi-major axis
e = eccentricity
i = inclination
P = period (in years)
T = epoch of periastron
Ω = position angle of the ascending node
ω = argument of periastron.

The relationship between the apparent and true orbits and the elements Ω, ω and i are illustrated in Figure 11.1.

Figure 11.2 is a diagram of the apparent orbit of Castor, one of the Gemini twins, with the conjugate diameters of the apparent ellipse drawn in. We construct the conjugate diameters by drawing the line acb from the center of the ellipse through the primary star to the periastron point on the apparent orbit. We then draw lines parallel to this line, tangent to the ellipse at points d and e. Finally, we draw a line through the center of the ellipse connecting d and e. The lines acb and dce are the conjugate diameters. The true orbit is the ellipse that results from

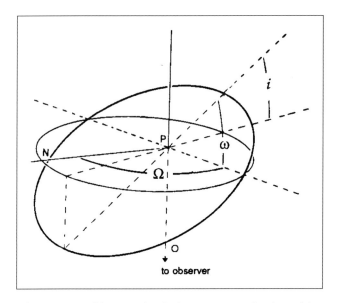

Figure 11.1. The projection of the true orbit of a binary star onto the plane of the apparent orbit.

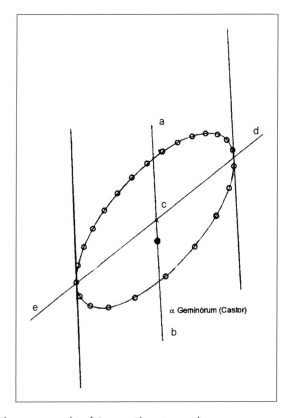

Figure 11.2. The apparent orbit of Castor with conjugate diameters.

projecting the apparent orbit, by methods of projection geometry, to a surface such that the conjugate diameters are mutually perpendicular.

The device shown in Figure 11.3 is designed to do this by visual rather than analytical projection.

It consists of a circular projection surface with two mutually perpendicular axes of rotation. Attached to the surface is a white screen with horizontal and vertical lines crossing at the center. The surface is between the tines of a fork made of 1/8 aluminum bar stock. It is important that the surface of the screen coincide with the center of rotation of the horizontal axis. The vertical axis of the fork is fastened to a 2 × 4 block so the center of the screen coincides with the center of the rectangular projection window. The vertical axis of the fork consists of a ¼ × 20 bolt fitted with a wing nut that locks it in place. The projection window, made of ¼-inch plywood with a rectangular opening, is mounted at

Figure 11.3. A binary orbit projector.

a distance slightly more than the radius of the screen from the center of the fork. A transparent copy of a binary star apparent orbit including its conjugate diameters is attached to this window with masking tape.

For the projector illustrated in Figure 11.3, the projection screen is 5.75 inches in diameter. The projection window frame is 8.5 × 10.5 inches and the window is 5 × 5.25 inches, but the dimensions are not critical. The most important construction detail is for the surface of the screen to coincide with the axis of the fork.

The orbit is projected with a high intensity desk lamp with a transparent bulb. Figure 11.4 illustrates the process. Using masking tape, attach the orbit transparency to the window so the major axis of the orbit coincides with the horizontal line on the screen and the intersection of the conjugate diameters coincides with the center of the screen. Then place the device in position about ½ meter or more in front of the lamp and rotate the screen on both axes until the projected conjugate diameters of the apparent orbit coincide with the perpendicular axes on the screen. The major axis of the transparency must coincide with the horizontal axis of the screen. When coincidence is achieved, lock the screen in place and you will see a representation of the orientation and shape of the true orbit in space relative to a plane perpendicular the line of sight as illustrated in Figure 11.5

For some orbits, such as Castor's, the projection of the major axis of the true orbit onto the plane of the sky is shorter than the orbit's minor axis. This is due to the inclination of the orbit about the minor axis. A line connecting the center of the ellipse through the primary star to the periastron point on the apparent orbit is always the major axis of the orbit.

The binary star orbit projector can satisfy your curiosity about what the true orbits of the binary stars are like. It is especially useful as a demonstration device for science educators.

Figure 11.4. Projecting a binary star orbit.

Figure 11.5. A projection of the apparent orbit onto the plane of the true orbit.

Figures 11.6 and 11.7 are diagrams of known binary star apparent orbits with conjugate diameters that can be scanned to transparencies for use with the projector. A computer program written in Basic for calculating position angles and separations

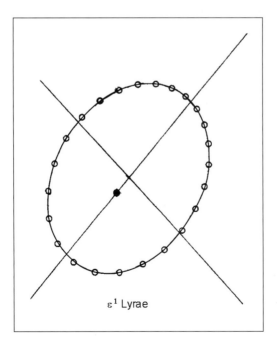

Figure 11.6. The apparent orbit of ε^1 Lyrae.

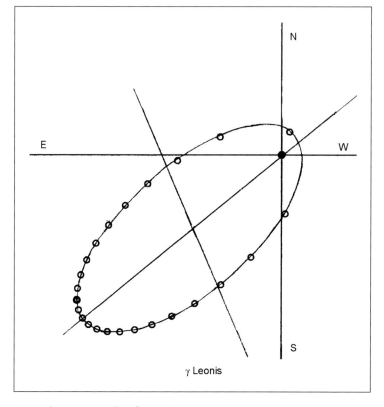

Figure 11.7. The apparent orbit of γ Leonis.

from the elements of binary star orbits can be found on page 120 of the 20th edition of *Norton's Star Atlas*. A CD containing software for determining separations and position angles for any date is included in *Observing and Measuring Visual Double Stars* edited by Bob Argyle. These sources can be used to plot additional apparent orbits. The projector works best for orbits with low eccentricity.

CHAPTER TWELVE

Visual Observations of Variable Stars

The brightness of many stars varies over periods ranging from hours to years. While some of this variation is due to the eclipsing activity of binary stars, other causes are intrinsic to the physical nature of the stars, themselves, as they evolve. Some stars vary their brightness with highly regular periodicity, others erratically and unpredictably. For many of these stars we have, at best, a statistically reliable database that spans little more than 100 years. Consequently, for a star with an inconsistent period of variability in excess of a year, we have less than 100 representations of its light curve from maximum to maximum. This is a small sample for a star that is passing through an evolutionary transition.

What we now know about variable stars is due in large part to the assiduous observations of amateur astronomers over the past century. They have provided a database from which professional astronomers draw for further investigation. Future astronomers will do the same with the data gathered today.

Although contemporary observations of many types of stars can best be made with large telescopes using sophisticated photometers with great precision, there are some types of bright variables for which using big telescopes is a waste of time and money. It makes little sense to measure a sixth magnitude star that varies 5 magnitudes to a precision of 0.01 magnitudes with an-800 mm telescope when for the purpose of establishing the periodicity and amplitude of the light curve, an experienced observer with an 80-mm telescope can measure it visually to an accuracy of 0.1 to 0.2 magnitudes. The accumulation of data from a large number of such observers gives a statistically reliable light curve for the star.

One type of variable star particularly suited for observers with small telescopes is the long period Mira, in that for many of them, much of their light curve lies within the range of an 80-mm refractor. The range of brightness for a particular long period variable can be as much as 10 magnitudes. Their periods range from a few months to more than a year. They are named after the prototype omicron Ceti, discovered to be variable in the seventeenth century. It was given the name Mira, meaning wonderful, in recognition of its remarkable ability to disappear, reappear and disappear again with a period of over 300 days.

Mira-type variables are all red giant stars believed to be evolving toward becoming planetary nebulae. The amplitudes of their magnitude change, not being constant may differ by as much as three or four magnitudes between adjacent maxima. Their periods may vary by as much as 15% from cycle to cycle. A few have shown steady decreases in their periods over the past century.

The Telescope

We measure variable star magnitudes visually by comparing them with stars of known magnitude. Consequently, one needs a wide field of view so that reference stars can be acquired with little or no movement of the telescope. Any telescope will do but long focal length types are more difficult to use because of their restricted fields. Short focal length telescopes with low power eyepieces and good edge of the field definition are best. If you are using a long focal length it would be useful to replace the usual 30-mm finder with one with a 50-mm aperture. This will make locating variable stars much easier.

Preparation for Observation

Beginning observers should make a list of a few easy-to-find long period variable stars (see Table 12.1) that will be approaching maximum brightness during their observation. For 80-mm refractors, the stars should have maxima brighter than eighth magnitude. When these stars approach maximum brightness their reddish color becomes obvious and they usually stand out among other stars in the field. Lists of variable star maxima for each month are published in *Sky and Telescope*, *Astronomy Now* and other astronomy periodicals. The AAVSO annually publishes Bulletin 68, a comprehensive list of predictions of long period variable star maxima. The 2005 edition of the RASC *Observer's Handbook* includes a list of 62 long period variables with maxima brighter than eighth magnitude.

After you've decided on the stars you want to observe, you need to find them, as well as suitable reference stars. For this you will need charts. With the right software you can make your own using Guide 8.0, an excellent program. It provides for making charts that match the field of view of your telescope, labeling the variables and magnitudes of reference stars, and setting the limiting magnitude level to correspond to that of your telescope. Standard charts can be downloaded from the AAVSO and similar organizations' websites. If you are using a refractor with a right angle prism or a Maksutov–Cassegrain telescope,

Visual Observations of Variable Stars

Table 12.1. Some Long Period Variable Stars

Star	RA	Decl.	Max.
R And	00h24m	+38°	5.8
S Scl	00h15m	−32°	5.3
U Ori	00h56m	+20°	4.8
O Cet	02h19m	−03°	2.0
R Vir	12h39m	+07⁰	6.1
T UMa	12h31m	+60°	7.7
S UMa	12h39m	+61°	7.8
S Vir	13h27m	+06°	7.0
R Hya	13h49m	−23°17′	3.5
R Cen	14h17m	−59°55′	5.3
R Nor	15h36m	−49°30′	5.0
X Oph	18h38m	+08°50′	5.9
R Aql	19h01m	+08°14′	6.1
T Cep	21h08m	+68°29′	5.2
R Aqr	23h38m	+15°17′	6.5
R Cas	23h58m	+51°24′	4.7

keep in mind when making or choosing a chart that east and west will be reversed in the field of view.

Charts made from Guide 8.0 and labeled with AAVSO standard chart magnitudes for a few Mira variables are included as Figures 12.1 through 12.4. The

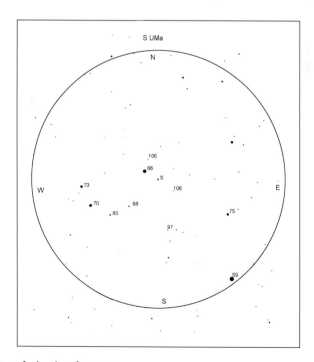

Figure 12.1. A finder chart for S UMa.

112 Real Astronomy with Small Telescopes

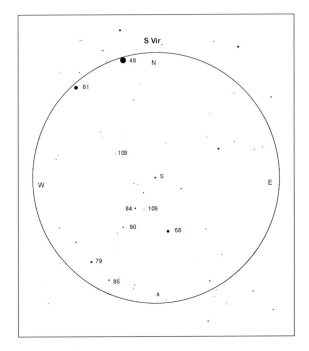

Figure 12.2. A finder chart for S Vir.

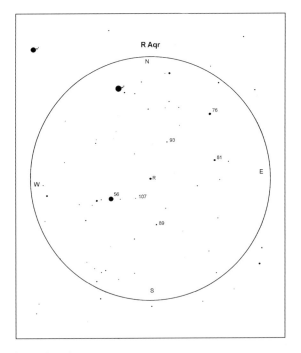

Figure 12.3. A finder chart for R Aqr.

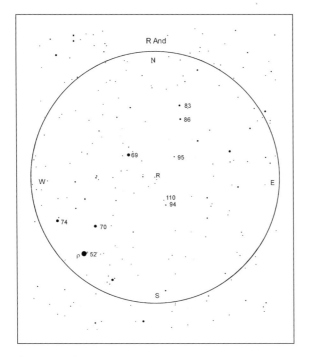

Figure 12.4. A finder chart for R And.

circle on the charts represents a 2°.5 field. Magnitudes are listed without the decimal point in order to avoid confusion with stars. For example, 1065 indicates a magnitude of 10.65. Before beginning your observation make a list of the right ascensions and declinations of the brightest stars near each of the variables in your program for the night. This will be valuable for locating each of the stars.

Making the Observation

When making the observation be sure your telescope mount is properly aligned with the celestial pole. Sometimes the variable star is near a bright star at nearly the same right ascension or declination. In that case you can simply move the telescope in right ascension or declination until you see a pattern of stars that you recognize as being similar to the field of your chart. Then look for the variable star. Otherwise you will need to use the setting circles. Although the circles on the EQ2 and similar small telescope mounts are small, they are good enough if you've made the modifications suggested in Chapter 2. Place the brightest star near your target star in the center of the eyepiece and note its right ascension and declination, as well as the coordinates of the target variable. Move the telescope through the difference in right ascension as read on the circle between the bright star and the variable. Do the same for declination. If you have done this carefully, the variable star will be within the 2°.5 field of view of a 25-mm eyepiece for an f/5 refractor. If you do not see a field similar to your chart, move the telescope slightly until you see a recognizable pattern. I find it best to turn the telescope drive motor off and move the telescope manually.

After you are certain you've identified the variable star look for the reference stars close to the same brightness. You may have to move the telescope field of view a little. Try to find three reference stars. If possible, find at least one that is brighter and another that is fainter than the variable. By comparing it to the magnitudes of reference stars estimate the magnitude of the variable.

Select a comparison star and arrange the telescope diagonal so that it and the variable star are horizontally aligned. Concentrating your vision on the variable and the comparison star for extended periods will make red variables appear deceptively brighter. Stars at the edge of the field will inevitably be slightly distorted or dimmed by optical effects. Move them toward center of the field for accurate comparisons. Look just long enough to make the comparison. It is better to take a series of quick looks rather than one long stare. Don't attempt to estimate the magnitudes of stars close to the limiting magnitude of your instrument as the measurement will be unreliable. For example, if the limiting magnitude of your telescope is 11.3, restrict your observations to stars no fainter than 10 magnitudes.

Note the name or designation of the variable star, the Universal Time and date of your observation, your magnitude estimate and the magnitudes of the reference stars. Also make note of any conditions such as moonlight, ambient light, haze and poor seeing that may have affected your observation.

The UT and date of your observation must be converted to the Julian day. A Julian day calendar can be downloaded from the AAVSO and similar organizational websites. It can also be obtained from Guide 8.0 software by going into the [Time Setting] menu.

After you have made the observation verify the accuracy of your estimate by going to the AAVSO website's Quick Look file, where you can compare it with others made the same night. Don't be prejudiced by others' results. Although yours may be a few tenths of a magnitude different from theirs, remember that the purpose of gathering data from many observers is to obtain a statistical database for a light curve of the star. Remember that all reported observations are evaluated for their reliability. If you are off by a magnitude or more, you have probably measured the wrong star. This can easily happen when you are not yet familiar with the star field or the process. Don't be discouraged. Try again another night. Strive for an accuracy of at least 0.2 magnitudes of the mean reported value.

When you have become confident with your ability to measure variable stars, consider joining the AAVSO or a similar organization and begin submitting observations. They will provide you with a more comprehensive outline of procedure and report forms. You will also be able to submit your observations online. They will then become part of the international database.

Restrict yourself initially to a few bright variables with which you are familiar and which you can easily locate. It is better to make four or five good measurements during an observing session than to spend the entire night searching for new stars and not measuring any of them. This is particularly true if your time is limited. You should have a list of favorites for each season.

After you have had some experience with long period variables you may want to expand your program to other types of variable star within the capabilities of your equipment. These may include eclipsing binaries and eruptive and irregular variables. A more complete treatment of the considerable breadth of variable star investigations is given in the sources at the end of the book.

CHAPTER THIRTEEN

Photography of Variable Stars

Accurate measurements of long period variables can be obtained with a digital camera. Digital photographs also have the advantage of recording a variable star when it is fainter than the visual limit of your telescope. A 30-second photographic exposure with an 80-mm refractor can dependably detect stars as faint as 11.5 magnitudes whereas the reliable limit for visual observations with that aperture is tenth magnitude. The camera also provides a permanent visual record of the star's light curve.

To take a digital camera photo of the field around a variable star use your lowest power eyepiece with the largest apparent field of view. I use a 25-mm Plossl eyepiece with a 52° apparent field on an 80-mm f/5 refractor. Be sure you have turned the telescope drive motor on, focused it and have centered the target star in the eyepiece. Lock the focusing device, remove the eyepiece and attach it to camera. Set the camera for ISO 400 and the longest exposure time that is possible without introducing noise. Carefully place the eyepiece with camera attached back in the telescope. Use the remote shutter release or the self-timer depending on which your camera has. You may not be able to see any stars in the camera's viewing screen. On the other hand if the telescope is rigidly mounted and you have been careful not to move it, the large field of view should include the variable and reference stars.

Processing the Image

When printing the digital image of a variable star field don't computer enhance either brightness or contrast as that will alter the magnitude vs. image brightness linearity. Fainter stars will be enhanced more than brighter ones. Print the picture and label it on the back with all of the same information you used to report your visual observation. Save the photo to the disk on which you archive all your variable star photos. Eventually you will have a permanent photographic record of long period variables to use for future reference.

The Method of Measurement

This method is based on the linear relationship between the diameter of a star image and its magnitude. Figure 13.1 is a calibration scale made from a circle template of the sort that can be found in any store that sells drafting supplies. The circles decrease in diameter linearly in steps of 1/64, from 5/16 to 1/16 of an inch.

I made this scale by placing the template on heavy white paper and filling in the circles with black ink. Then I scanned it to a computer and converted it to white images on a black background. The length of the scale is suitably adjusted when printed by using the [Image Size] button on your photo processor. Number the scale images starting 1.0.

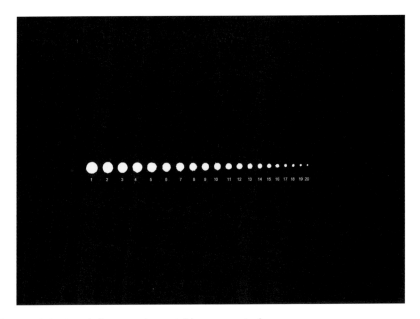

Figure 13.1. A scale for measuring variable star magnitudes.

If you wish, you can save yourself all that work by scanning the scale in the book. Save the scanned image on your variable star archive disk for later use.

To use the scale, prepare the print by converting it to black and white and then cropping it to enlarge the star images so they match the range of image diameters on the scale. In the cropped print include a minimum of four suitable reference stars, excluding the outer regions of the print where stars are distorted due to lens aberrations. You will usually include the central 1.5° diameter of the print.

The scale images are, in effect, artificial star images to which the reference stars and the variable will be compared. The numbers assigned to them are not magnitudes but a dimensionless measurement reference. The photographic images will not match the full range of the scale. This is especially true if all the reference stars are relatively faint. Nevertheless, the full range of the reference stars should match some range of images on the scale.

The scale should be cut into a thin strip so that a scale image can be placed adjacent to the image of the star being measured. Place the scale image horizontally rather than vertically next to the star.

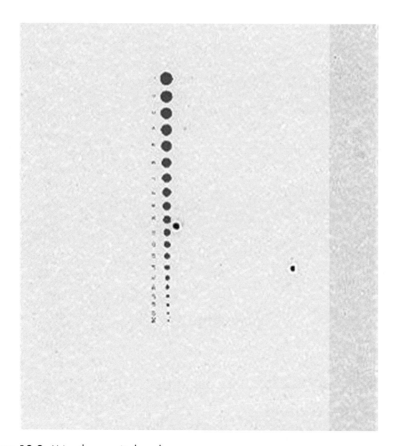

Figure 13.2. Using the magnitude scale.

Making the Measurements

To make a measurement place the scale next to a reference star, as shown in Figure 13.2, and estimate which scale image most nearly matches the star image. If you estimate that it's between the values of two scale images, add 0.5 to the brighter of the two. Make measurements of as large a range of reference stars as you can in the area around the variable star, as well as the variable. Do not include any images that may be elongated due to optical aberrations. Using either an AAVSO or comparable standard chart, record the magnitude of each reference star you measure.

Analyzing the Data

Using rectangular coordinate paper, plot reference star magnitudes vs. image scale numbers and draw a best-fit straight line through the points. Plot the value of the scale reading for the variable star on the line to determine its magnitude for the observation. If you have a scientific pocket calculator with a linear regression mode, you can enter the data on the calculator and get the value of the variable star magnitude directly.

Figure 13.3 is a plot of data taken from a digital photo of Chi Cygni on JD 2453647.7. The photograph was taken using a 5-megapixel camera with an 80-mm f/5 refractor. The mean value from measurements of three different photographs taken on the same night was 9.0 with a standard deviation of 0.08 magnitudes. The linear correlation for each of the plots is approximately 1.0. An AAVSO

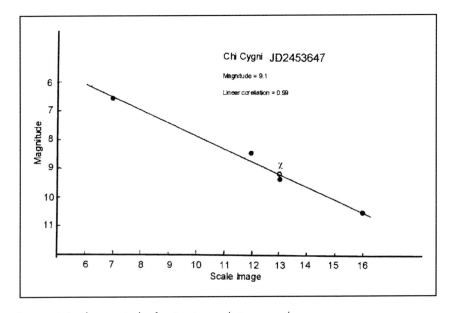

Figure 13.3. The magnitude of χ Cygni vs. scale image number.

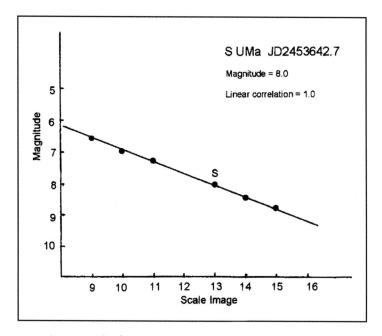

Figure 13.4. The magnitude of S UMa vs. scale image number.

chart was used for reference stars and a linear regression computation was used to determine the magnitude of Chi Cygni from each of the prints.

The mean value for measurements reported in the AAVSO Quick Look file for that date was 9.1 with a standard deviation of 0.09 magnitudes. This value included those of three observers using standard visual methods. For six observers reporting their measurements on JD 2453646 and 2453647, the result was 8.98 with a standard deviation of 0.33 magnitudes.

Figure 13.4 is a plot for the variable S UMa on September 29, 2005 (JD 2453642.7). A linear regression analysis with five reference stars gives a value of 8.0 for the magnitude of S UMa on that date with a linear correlation of 0.98. The values reported in the AAVSO Quick Look file for visual measurements of the magnitude of S UMa were as listed in Table 13.1. These have a mean value of 8.05 with a standard deviation of 0.17.

For photographic measurements of variable stars, digital cameras have some significant advantages over film. Their linear response and greater sensitivity

Table 13.1. Reported Magnitudes for S UMa

Julian Date	Magnitude
2453642	7.8
2453643	8.0
2453645	8.2
2453645	8.2

make it possible to record fainter stars in a shorter period of time. By using the method described here, they produce results as accurate and consistent as the best direct visual observations.

Film Photography

A plot of magnitude vs. star image diameter is not linear for photographic emulsions. In addition, the shape of the response curve differs with different films. Consequently, the method of analysis described here for digital camera images is not applicable to film photography.

Magnitude estimates, however, can be made from a film print in the same way that visual estimates are made at the eyepiece. In addition, film photography has the advantage of being able to reach stars fainter than the visual limit of a small telescope. A two-minute exposure on ISO 400 film with a 200-mm telephoto lens will reach stars of 13th magnitude.

Black-and-white film has extended blue sensitivity but Mira variables are red. Consequently the magnitudes obtained from film estimates are different from visual observations that are red biased. One can get a close approximation to visual magnitudes by using a yellow filter. Any report of magnitudes measured on film should indicate they were made photographically and whether or not a filter was used.

CHAPTER FOURTEEN

Star Clusters and Nebulae

The general opinion that a short focus low power instrument is best for observing nebulae and star clusters is not necessarily true. The best combination of field of view and magnification depends on the particular object being observed. For instance, the double cluster in Perseus (Figure 5.1) is spectacular in a 25-mm eyepiece used with an 80-mm f/5 telescope. On the other hand, globular clusters and planetary nebulae are better seen at 100× with an 80-mm f/11 refractor. The contrast between a faint object and the sky background is less at f/5 than at f/11 for the same magnification. A low f/ratio instrument is advantageous for wide field views of the Milky Way, extended star clusters and digital photography of those objects.

In reality the best instrument is the one you have. Any of the small telescope types we have considered here should be capable of delivering interesting views of the brighter star clusters, nebula and galaxies. You will need dark, clear, moonless nights.

In 1784 Charles Messier published a list of 103 star clusters and nebulae to provide a catalog of known objects that could be mistaken for comets. Until early in the twentieth century all visual records of these objects were made by astronomers patiently reproducing what they observed in sketches.

Figure 14.1 is a drawing of the great nebula in Orion (M41) done by W. C. Bond, the first director of Harvard observatory, in the middle of the nineteenth century. It is the result of hours of observation of the nebula with the 15-inch Harvard

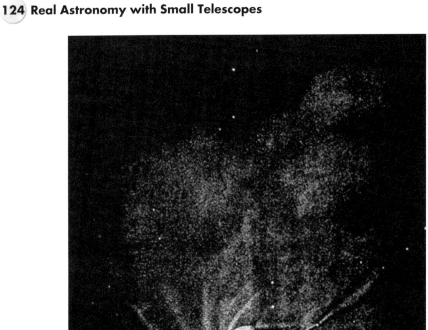

Figure 14.1. M42 in Orion as sketched by W.C. Bond.

refractor at Cambridge, which shared with the Pulkova telescope in Russia the honor of being the world's largest refractor.

It is interesting to see how views through a modern small version compare with sketches done by a great nineteenth-century observer with what was once

Table 14.1. Messier Objects

Designation	RA	Decl.	Magnitude	Description
M 31	$0^h 42^m$	+41°16′	3.4	Galaxy in Andromeda
M 1	$5^h 35^m$	+22°01′	8.4	Crab nebula in aurus
M 42	$5^h 35^m$	−5°27′		Orion nebula
M 81	$9^h 56^m$	+69°04′	6.9	Spiral galaxy
M 82	$9^h 56^m$	+69°41′	8.4	Galaxy
M 51	$13^h 13^m$	+47°12′	8.4	Whirlpool galaxy
M 101	$14^h 03^m$	+54°21	7.9	Pinwheel galaxy
M 57	$18^h 54^m$	+33°02′	8.8	Ring nebula in Lyrae
M 27	$19^h 59.6^m$	+22°43′	7.3	Dumbell nebula

one of the world's largest telescopes. Table 14.1 lists some of the brighter Messier objects that can be observed with small telescopes.

Digital Photography of Star Clusters

Because nebulae and galaxies are extended diffuse objects, with the exception of the bright nebula M42, the great nebula in Orion, digital photography with small telescopes will not reveal more than the eye can see. Common fixed lens digital cameras do not have sufficient noise reduction for the long exposures needed. Although digital single lens reflex cameras are available with ISO speeds of 800 or more, resolution of 6 to 8 megapixels and much longer low noise exposure times, we consider them too expensive to be used within the context of this book's intentions.

Star clusters, however, consist of numerous point sources and a 30-second exposure with a fixed lens digital camera will reveal much more than the eye can see in the telescope. Figure 14.2, for example, is a 30-second exposure of the globular cluster M13 in Hercules obtained with an 80-mm f/6 refractor. Visually, in the telescope this cluster appears as a luminous, circular patch of light with no distinguishing features.

Stars clusters, in general, are of two types – globular and galactic. Globular clusters are spherical aggregates of 10,000 or more stars that lie in a halo surrounding the plane of the Milky Way. They contain the oldest stars in our Galaxy. Galactic clusters are loosely associated arrays of stars that formed in the plane of the Galaxy. Table 14.2 lists some star clusters that are good objects for small telescope digital photography.

For extended exposure photography of nebulae and clusters, a short focal length telescope (f/5) is preferred and accurate alignment of the mount with the celestial pole is essential. This will make 30-second exposures without elongating the star images. Use the same procedure described in Chapter 9 for variable star photography.

If the contrast is increased, downloaded photos will show fainter stars. For these objects, preserving the linearity of the camera response is not a concern.

126 Real Astronomy with Small Telescopes

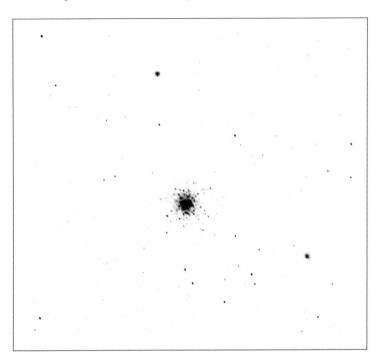

Figure 14.2. A digital photo of the globular cluster M 13 taken with an 80-mm f/5 refractor.

Table 14.2. Star Clusters for Digital Photography

Designation	RA	Decl.	Magnitude	Description
M 13	16^h42^m	+36°26'	5.9	Globular Cluster
M 12	16^h47^m	−1°57'	6.6	Globular Cluster
M 10	16^h57^m	−4°06'	6.6	Globular Cluster
M 92	17^h17^m	+43°08'	6.5	Globular Cluster
M 15	21^h30^m	+12°10'	6.3	Globular Cluster
2802 Car	09^h12^m	−64°52'	6.3	Globular Cluster
3201 Vel	10^h18^m	−46°25'	6.7	Globular Cluster
ω Centauri	13^h27^m	−47°29'	3.7	Globular Cluster
NGC 884, 869	02^h20^m	+57°08'	4.4	Double Cluster in Perseus
M 34	02^h42^m	+42°47'	5.2	Open Cluster
M 45	03^h47^m	+24°07	1.2	Pleiades
M 38	05^h50^m	+32°08'	6.4	Open Cluster
M 36	05^h36^m	+34°08'	6.0	Open Cluster
M 37	05^h52^m	+32°33	5.6	Open Cluster
M 44	08^h40^m	+20°00'		Beehive Cluster in Cancer
M 29	20^h24^m	+38°32'	6.6	Open Cluster
M 39	21^h32^m	+48°26'	4.6	Open Cluster
M 52	23^h24^m	+61°35'	6.9	Open Cluster
4755 Cru	12^h54^m	−60°20'	4.2	Jewel Box
M 6	17^h40^m	−32°13'	4.2	Butterfly Cluster
M 7	17^h54^m	−34°49'	3.3	Open Cluster

Increasing contrast too much, however, will introduce a significant amount of electronic noise.

W. C. Bond and his son G. P. Bond obtained the first Daguerreotype image of a star in the 1850s. They needed a 20-minute exposure to record Vega with the Cambridge 380-mm refractor. Compare this with a 15-second exposure of a star cluster using your digital camera on a small telescope.

CHAPTER FIFTEEN

 # A Color–Magnitude Diagram for The Pleiades

As a source of radiant energy gets hotter, it gets brighter and its color shifts toward the blue end of the spectrum. Red stars have surface temperatures of about 4000 K; blue ones exceed 15,000 K. The quantitative measure of a star's color then is an indication of its temperature, is the color index. It is derived from measurements of the difference between the apparent magnitudes in two different regions of the spectrum.

A common method is to measure the magnitude of a star with a photometer in which filters are used to limit the measurement to different regions of the spectrum. When the measurement is limited to the blue region it is called the B magnitude. A filter that matches the response of the human eye to color is the V magnitude. The color index is then given by

$$CI = B - V$$

The absolute magnitude is a measure of the intrinsic luminosity of a star. By comparing absolute magnitudes we are comparing them as if they were all at the same standard distance of 10 parsecs. Consequently, stars of lower absolute magnitude number are (for example, $M = -2$) intrinsically more energetic than stars of higher magnitude number (for example, $M = +5$). We should also expect to find the intrinsically brighter stars to have higher temperatures and bluer colors. A plot of absolute magnitude vs. color index for a population of stars is called a color–magnitude diagram. If a particular group of stars all belong to the same star cluster, absolute magnitudes need not be determined. Since they

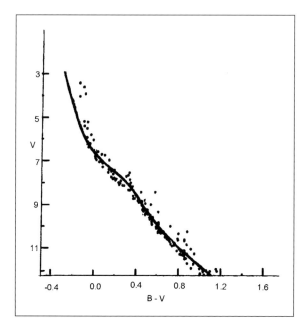

Figure 15.1. A color–magnitude diagram for a young star cluster.

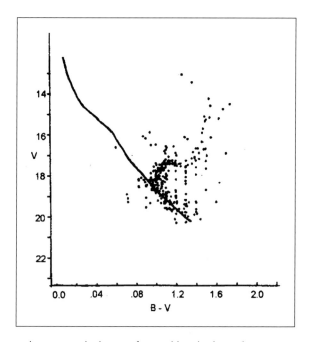

Figure 15.2. A color–magnitude diagram for an old evolved star cluster.

A Color–Magnitude Diagram for The Pleiades

are all at approximately the same distance, their apparent magnitudes and color indexes can be compared. This can be done for some of the brighter relatively nearby star clusters by measuring prints from a digital camera attached to a small telescope.

Figure 15.1 is a color–magnitude diagram for a young star cluster, the stars of which occupy a narrow band called the main sequence. The curve running through the distribution defining the zero-age main sequence represents the point in the star's evolution at which nuclear fusion begins. In Figure 15.2 we see a much older cluster in which many of the main sequence stars have evolved along a horizontal giant branch; some to the red super giant stage.

Acquiring the Data

The data consist of measurements from three different 30-second digital camera images of the Pleiades cluster in Taurus. Take the first one without a filter. Repeat the procedure described for variable stars in Chapter 12. Take several photos so you can choose the best for measurement. Follow these pictures with a series taken with a light-blue filter in the eyepiece and then another with a light-yellow filter. With the exception of filters be sure to keep all the photographic parameters constant throughout. Download the best image from each series and make black-and-white glossy prints. Figure 15.3 is an example of what the photographs should look like.

Print a negative of one of the photos on plain paper to provide a reference for numbering and recording the stars to be measured, and randomly select 15 to 20 with a distribution from the brightest to the faintest around the center. Make certain you don't use star images distorted by lens aberrations.

Figure 15.3. A digital camera photo of the Pleiades taken with an 80-mm f/5 refractor.

The Analysis

For measuring these prints use the magnitude scale and the variable star procedure in Chapter 13. Place the scale horizontally adjacent to the star to be measured. Then compare the star with the scale images to determine the best match. Remember the scale readings are not magnitudes but numbers proportional to the magnitudes of the stars. From the negative print, pick out one of the numbered stars and measure it on both blue- and yellow-filtered as well as the unfiltered prints. Designate the scale readings from the blue-filtered prints as B and those from the yellow as Y. Designate the readings from the unfiltered prints as m_o. Understand these are not standard magnitudes but they could be converted by comparing them to standard B and V magnitudes. You don't have to do this, however, to get the shape of the color–magnitude distribution.

The difference between the scale readings for the same star on the two different prints is a color index for that star. We can call it a B–Y index. A plot on rectangular coordinate paper of m_o vs. B–Y will be a color–magnitude diagram for the cluster.

As a star cluster ages, more of its stars become red giants and the color–magnitude develops a horizontal branch. What does your color–magnitude of the Pleiades tell you about the age of that cluster? After you have done a color–magnitude diagram for the Pleiades, you might try other clusters such as the Beehive in Cancer and the Hyades in Taurus.

CHAPTER SIXTEEN

The Design of an Objective Prism Spectrograph

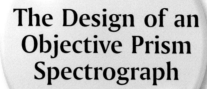

Stellar spectra are classified into several distinct types according to the prominence of certain absorption or emission lines. This classification is also an indication of a star's surface temperature and color. The most common types are classified according to the labels O, B, A, F, G, K and M, with each type divided into 10 subclasses such as A0, A1, A2, A3, etc. Since they are very rare types W, R, N and S are not presented here. Nearly 99% of all stars are of types B to M. Table 16.1 is an outline of the classification of stellar spectra.

Figure 16.1 illustrates the objective prism method for obtaining stellar spectra of several stars simultaneously.

The angle through which red rays of light will be refracted by a prism is given by

$$\Theta_r = 2\arcsin[N_r \sin(A/2)] - A$$

Where A is the prism angle and N_r is the index of refraction of the prism for a particular wavelength of red light. For violet rays, this becomes

$$\Theta_v = 2\arcsin[N_v \sin(A/2)] - A$$

If F is the focal length of the objective lens, the linear dispersion D, the length of the spectrum at the film plane, is given by

$$D = F(\Theta_v - \Theta_r)\pi/180°$$

Table 16.1. The Classification of Stellar Spectra

Type O	Ionized helium, neutral hydrogen, and helium absorption lines, blue in color, with temperatures of 50,000 K or higher
Type B	Weak hydrogen Balmer lines, temperatures around 25,000 K
Type A	Hydrogen Balmer lines strongest. H and K lines of CaII appear in later types. White stars with a temperature of 10,000 K
Type F	Hydrogen Balmer lines weaken, K line of CaII becomes prominent. Yellowish white in color with a temperature of 7,500 K
Type G	Balmer lines weaker than type F. H and K lines of CaII are stronger; many lines due to neutral metallic elements. Yellow stars with temperatures around 6,000 K. The Sun is a type G2 star
Type K	Lines due to neutral metals become prominent, hydrogen lines are very faint. These stars are orange, with temperatures around 4,500 K
Type M	Strong lines of neutral metals. Strong titanium oxide bands. Red stars with temperatures around 3,500 K

Large telescopes use objective prisms with angles of only a few degrees and a long focal length. Equivalent results can be obtained with a large prism angle and a short focal length. Although a 60° equilateral prism would be impossibly heavy and difficult to mount on a large telescope, it is perfect for use with a 135-mm telephoto lens on a 35-mm SLR. For a flint glass prism the linear dispersion is 13.4 mm at the focal plane of a 135-mm telephoto lens.

Figure 16.2 illustrates the basics of the design for an objective prism spectrograph on a 35-mm camera. Equilateral prisms with a 40-mm face can be purchased from Edmund Scientific Company. The prism face should be large enough to nearly cover the diameter of the camera lens. If it doesn't, the open areas of the lens can be masked off. Its inclination relative to the camera lens should be such that the face of the prism farthest from the lens is parallel to it.

The objective prism spectrograph described here can capture good spectra of first- and second-magnitude stars. Stars of all of the major spectral classes are within this range. Since the planets Jupiter and Saturn shine by reflected sunlight, they are good sources for photographs of the solar spectrum because their images are nearly point sources on the scale of a 135-mm lens.

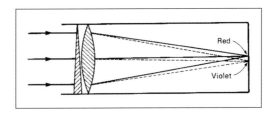

Figure 16.1. The objective prism method of obtaining stellar spectra.

The Design of an Objective Prism Spectrograph

Figure 16.2. An objective prism spectrograph.

For stellar spectra, black-and-white film produces greater image contrast and resolution of absorption lines. Color film spectra show dark bands where different primary color emulsions overlap. Digital cameras do the same. The overlapping dark bands manifest themselves when color is dispersed by a prism. A digital camera can be operated in a black-and-white mode but the overlapping of the different color sensitivities will still persist in the image of a spectrum.

Getting the Spectrum

A 35-mm SLR film is the only type of camera that can be used effectively for acquiring stellar spectra with an objective prism. If you don't already have one, maybe you can find an inexpensive used one. It needs to be capable of 2-minute exposure times and have a cable release for the shutter. Any lens will do. The standard 50-mm lens will provide enough dispersion to at least demonstrate the differences between spectral classes for bright stars; the longer the focal length the greater the dispersion.

The camera can be mounted piggyback on the telescope, as shown in Figure 16.3, or by itself on an equatorial. A polar axis drive is not necessary. The camera has to be arranged so that the prism's dispersion is in the declination direction and the lens points approximately 60° away from the target star. The best way to do this is to line up the telescope or finder (Figure 16.3) on the target star and then rotate the camera until the spectrum is visible in the viewing screen.

Figure 16.3. The spectrograph mounted on an 80-mm f/5 refractor.

The spectrum as seen in the camera is a dispersed line without width. It becomes widened through the process of letting the star drift across the field of view in the right ascension direction. A two-minute exposure will give the spectrum sufficient width with a 135-mm lens. Short focal lengths will require longer exposures. Use black-and-white film with an ISO of 400.

Figures 16.4 and 16.5 are spectra of the star Sirius and M42, taken with the spectrograph illustrated in Figure 16.2.

The spectrum of Sirius clearly shows the hydrogen Balmer series of absorption lines characteristic of spectral class A stars, with the red end of the spectrum to the right. The hydrogen α line in the far-red section is not visible because the film sensitivity (T Max 400) cuts off sharply at that wavelength. The first visible absorption line is the hydrogen β in the blue-green area. Although the human eye is insensitive to wave lengths much beyond the hydrogen δ, the third absorption line from the right, film sensitivity extends into the far violet.

In the Orion nebula's spectrum there are three emission lines caused by the excitation of the gases by ultraviolet light from its embedded young stars. These are bright lines in contrast to dark absorption lines characteristic of stellar spectra. The third emission line from the red end is the hydrogen β. The other two to the right of the H_β were of controversial origin when they were first discovered because their wavelengths did not match the spectra of any then known element. It was hypothesized they were caused by a previously unknown element characteristic of interstellar gas, which they designated nebulium.

With advances in quantum theory that could predict the rate at which possible energy transitions could take place, however, it was discovered that the nebulium lines were actually produced by doubly ionized oxygen. These lines are not visible because in the best laboratory vacuums, the rate at which those particular

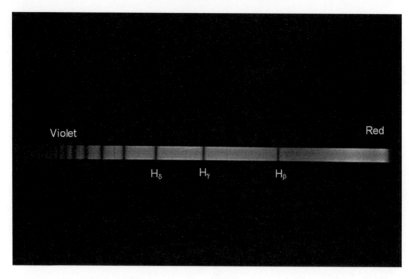

Figure 16.4. The spectrum of Sirius.

transitions take place is greater than the time between collisions of atoms. The density of the Orion nebula is so low, however, that the interval between collisions is much longer. The oxygen atoms are essentially isolated. As a result the low probability transitions can be observed.

Outer space expands the limits of experimental physics. In that vast laboratory we can observe the behavior of matter at higher and lower pressures, higher and

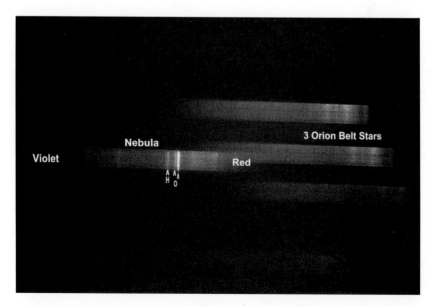

Figure 16.5. The emission spectrum of M 42.

lower densities, and higher and lower temperatures under the influence of much greater gravitational and electromagnetic fields than is possible in any facility we might construct on Earth. Thus we gather evidence necessary to support or reject theories of the fundamental properties of matter and the interactions that govern the structure of the universe.

CHAPTER SEVENTEEN

The Proper Motion of Barnard's Star

Because all the stars in our Galaxy are in orbits around its center, nearby stars demonstrate motion relative to those more distant. This motion has two components. One of these, the motion toward or away from us in the line of sight is the radial velocity. The motion perpendicular to the line of sight, measured in units of arc seconds/year, is termed the proper motion of the star. The proper motion, itself, has components in right ascension and declination.

Barnard's star, a tenth magnitude object at a distance of 5.9 light years, has a proper motion in declination of 10.3 arc seconds/year. This is easily within the observational limits of a small telescope.

Barnard's star, 17^h 58^m RA and Decl. $+4°\,42'$, is within two degrees of the fourth magnitude star 66 Oph and 3.5° due east of β Oph. Figure 17.1 is a chart for a 2° field made from Guide 8.0. The position of Barnard's star is indicated by the arrow.

Taking the Photographs

In order to easily observe the changing position of Barnard's star, a large print scale and therefore a long focal length is necessary. An 80-mm f/11 refractor or 90-mm Maksutov is recommended for these photographs. Accurate polar alignment of the telescope is imperative. An 80-mm f/5 refractor can be used but it will take twice as long between photographs to detect appreciable motion in the star.

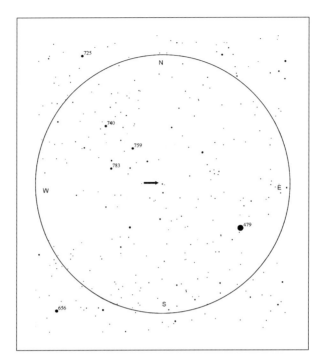

Figure 17.1. A finder chart for Barnard's star; the arrow indicates the star's position.

The afocal procedure for binary star photography is used here. With an 80-mm f/11 refractor, use a 6-mm wide field eyepiece such as Orion Expanse. A 9-mm eyepiece can be used with a 90-mm Maksutov. A 6-mm eyepiece combined with a 2× Barlow lens will work with an 80-mm f/5 refractor. The 90-mm Maksutov has the advantage of a short tube and fewer tracking problems. Since the change in the star's position is primarily in the declination direction, a slight amount of star trailing in the photograph will not obscure its proper motion.

Record the date, time and Julian day of your observation. Take several pictures and choose the best for printing. Before printing, crop the image to 100 mm × 100 mm to enlarge it. This will produce a print scale of about 1 arc second/mm. Save the image on a disk for later use.

Follow the first observation by another about a year later. In this period, Barnard's star will have moved about 10 arc seconds in declination. Use the same printing and cropping procedure you used for the previous photograph. The change in the position of the star should be on the order of 10 mm for the one-year period between the observations. The change can be seen easily by placing a transparent copy of the first observation over a print of the second.

A more quantitative measurement of the proper motion of Barnard's star can be obtained by printing each of the observations on a separate 1 mm × 1 mm grid of rectangular coordinate paper. By referring to the grid lines, measure the vertical distance between two widely separated stars of known declination and use it to establish a print scale in arc seconds/millimeter. Then measure the vertical distance between Barnard's star and some other star of known declination. Using this distance and the plate scale, calculate the declination of Barnard's star. Repeat the procedure for the print taken a year later. The difference between the two declinations divided by the Julian day difference will be the proper motion of Barnard's star in declination in arc seconds/year.

References and Additional Reading

Star Atlases

Tirion W, Sinnott R (2005) Sky atlas 2000, 2nd edn. Sky Publishing Corp., Cambridge.
Ridpath I (ed) (2004) Norton's star atlas, 20th edn. Pi Press, New York.

Telescopes and Accessories

Sidgwick B (1980) Amatuer astronomer handbook, 3rd edn. Dover Publications, New York.
Harrington PS (2002) Star ware, 3rd edn. Jossey-Bass, San Francisco.

Astrophotography

Covington M (2004) Astrophotography for the amatuer, 2nd edn. Cambridge University Press, Cambridge.
Reeves R (2005) Introduction to Digital Astrophotography. Willmann-Bell Inc., Richmond.
Monks N (2005) Astronomy with a home computer. Springer, Berlin Heidelberg New York.

References and Additional Reading

The Sun

Beck R, Hilbrecht H, Reinsch K, Volker P (1995) Solar Astronomy handbook. Willmann-Bell Inc., Richmond.

Taylor P (1991) Observing the Sun. Cambridge University Press, New York.

The Moon

Grego P (2005) The Moon and how to observe it. Springer, Berlin Heidelberg New York.

Rukl A (2004) Atlas of the Moon. Sky Publishing Corp., Cambridge.

Binary Stars

Argyle B (ed) (2004) Observing and measuring visual double stars. Springer, Berlin Heidelberg New York.

Variable Stars

North G (2004) Observing variable stars, novae and supernovae. Cambridge University Press, New York.

Levy DH (1989) Observing variable stars. Cambridge University Press, New York.

Star Clusters and Nebulae

Jones KG (1991) Messier's nebulae and star clusters. Cambridge University Press, Cambridge.

Index

AAVSO Bulletin 68, 110
Absolute magnitude, 129
Absorption spectra, 133
Accessories, 27, 143
Acquiring data for the Pleiades cluster C–M diagram, 131
Adapters for almost any camera model, 31
Adjust to ambient temperatures, 25
Afocal method, 29, 31, 35, 49, 59, 73, 93, 101
Airy disk, 22
ALPO, 42, 43, 75, 76
Antionadi, 62
Aphelion, 5
Apochromatic, 24, 28
Apparent magnitudes, 129, 131
Apparent orbit of Castor, 103, 104
Apparent solar day, 5
Apparent solar time, 5, 6
Asteroid orbit chart, 83
Asteroids, 13, 25, 30, 83
Astigmatism, 23

BAA, 42
Barnard's star, 139–141
Binary star apparent orbit, 106, 107

Binary star orbital elements, 100
Binary star position angle and separation, 92
Binary stars, 13, 25, 26, 28, 91–101, 103–108, 140
Binary star separation calibration stars, 96
Binary stars that are possibilities for 80-mm refractors and 90-mm Maksutovs, 96
Binary star true orbit projector, 103–108
Binary star true relative orbit, 91
Bond, W.C., 123

Camera should be small and light, 31
Celestial equator, 1, 3, 5, 7, 64
Celestial sphere, 1–3
Changing solar magnetic field, 37
Charts made from Guide 8.0 and labeled with AAVSO standard chart magnitudes, 111
Chopping block equatorial, 19

Classification of stellar spectra, 134
Color filters for planetary observations, 27
Color index, 91, 129, 131, 132
Color–magnitude diagram, 129, 130, 131
Color–magnitude diagram analysis, 130
Color–magnitude diagram for the Pleiades, 129–132
Coma, 23, 25, 79, 80, 81
Combination of film imaging and digital processing, 35
Comet dust tail, 79
Comet gas tail, 79
Comets, 25, 30, 79–89
Comets change their appearance, 79
Comets Hyakutake and Hale–Bopp, 80
Common point-and-shoot digital camera, 29
Comparison star, 81
Computer generated position of the asteroid, 83
Conjugate diameters of the apparent binary star orbit, 103

145

Index

Cooling-down problem, 26
Coordinates of ecliptic latitude run parallel to the ecliptic, 64
Craters Messier A and B, 56
Cropping also magnifies regions of interest, 34

Declination, 3, 7, 8, 9, 10, 11, 15, 18, 41, 64, 113, 135, 139, 140
Declination drive, 9
Detailed study of the surfaces of the planets, 61
Diffraction of light, 21, 22
Diffuse nebulae and galaxies, 30
Digital camera photo, 34, 52, 131
Digital camera photographs of the Moon, 52
Digital photography, 13, 25, 26, 29–31, 42, 62, 82, 83, 92, 125, 126
Digital photography of asteroids, 83
Digital photography of binary stars, 92–96
Digital photography of comets, 82
Digital photography of the planets, 62–64
Digital photography of star cluster and nebulae, 125
Digital photography of the sun, 42
Digital photography of variable stars, 117–122
Double binary star system epsilon Lyrae, 92
Double cluster in Perseus, 29, 30, 126
Drawing of the great nebula in Orion (M41) done by W.C. Bond, 123

Ecliptic, 1, 64, 69, 70, 74
Effect of poor seeing, 25
Effective focal length, 22, 31, 32
The elements of a binary star orbit, 108
Enhance the appearance of details in the atmospheres of Venus, Jupiter and Saturn, 62
Epoch, 7, 8
EQ1, 9

EQ2, 9, 10, 13, 19, 28, 113
EQ3, 9, 10, 13, 15, 28
Equation of time, 5, 6
Equatorial mount, 15
Equinox, 1, 2, 3, 7, 64, 71
Equinox of 2000.0, 7
For extended exposure photography of nebulae and clusters, a short focal length telescope (f/5) is preferred, 125
Extended noise-reduced exposure capability, 31
Eyepiece, 9, 15, 19, 20, 22, 26, 29, 30, 31, 32, 34, 39, 41, 51, 52, 62, 79, 81, 83, 93, 94, 99, 100, 113, 117, 131, 140

f/ratio, 22, 23, 24, 25, 30
Faculae, 38
Field curvature, 23
Field of view, 15, 19, 20, 25, 26, 28, 31–32, 41, 83, 110, 114, 117, 136
Filar micrometer, 92
Film photography of comet, 82
Film photography of the sun, 49
Filters that thread into standard eyepieces, 62
Finders, 27
First Daguerreotype image, 127
Fixed lens digital cameras, 29, 30, 125
Focusing device, 27, 117
Fork mount, 10, 19, 20, 28
Full aperture filters, 42

Galactic clusters, 125
λ Virginis, 100
GEM, 9, 11, 18, 19, 41, 95, 100, 101
Geocentric ecliptic longitude, 74
Globular clusters, 29, 123, 125
Go-to systems, 20
Greenwich meridian, 6, 8
Guide 8.0, 8, 53, 139

Height of a lunar mountain or crater wall, 53
Heliocentric longitude, 64, 71, 83
The heliographic longitude of the Sun's center, 43
Hershel, Sir William, 91
Hevelius, 51
Hyginus rill, 56

Increased solar activity, 38
Internal tube currents, 26

Julian day, 8, 78, 114, 140

Klee Barlow lens, 24

Latitude and longitude of sunspots, 42
Limitations, 21, 23, 25
Limitations in resolution, magnifying power and observable stellar magnitude, 21
Linear dispersion, 133, 134
Linear regression analysis, 121
Linear relationship between the diameter of a star image and its magnitude, 118
Linear scale, 34, 56, 57, 58, 75, 78, 92
Long focus refractors, 10
Longitude of the ascending node, 64
Long period variable star maxima, 110
Long period variable stars, 30, 110, 111
Lunar Alps, 56
Lunar Apennines, 56
Lunar coordinate grid, 58
Lunar libration, 57, 58
Lunar libration in longitude, 57
Lunar longitude of an object casting a shadow, 56
Lunar measurements on a computer monitor screen, 56
Lunar mountains, valleys, rills, faults, craters and plains, 51

Magnitude limits, 22
Magnitude vs. image size for Chi Cygni, 120
Magnitude vs. image size for S UMa, 121
Making a measurement with the variable star calibration scale, 118
Maksutov–Cassegrain, 9, 10, 19, 21, 25, 110
Maksutovs, 21, 39, 41, 96
Mars can only be observed during oppositions, 71
Maunder, E.W., 38
Maunder minimum, 38

Index

Mean solar time, 5
Measure objects directly on a computer monitor, 34
To measure the separation between the components of the binary star, 96
Measuring binary star photographs, 95
Measuring a series of digital photographs of Venus, 71
Messier, Charles, 123
Microsoft Picture It! Premium 10, 34
Minimum for the current solar cycle, 38
Minimum resolution of 3.1 megapixels, 31
Mira-type variables, 110

Newtonian reflectors, 25

Objective prism spectrograph, 133–138
Observing star clusters and nebulae, 123
Obstruction of light by the secondary mirror, 22
Obtaining objective prism spectra, 134
Older star cluster, 130
Open star clusters, 29
Orientation of the Sun's axis of rotation for different times, 43
Orion Expanse eyepieces, 26, 93
Orion nebula's spectrum, 136
Overlays, 34

Parabolic primary mirror, 25
Percival Lowell, 62
Performance of an 80-mm f/5 refractor, 24
Periastron, 100, 103, 106
Perihelion, 5, 64, 71, 73
Phases of Venus, 71
Photograph of the asteroid Vesta and the computer generated star chart for the date of the observation, 83
Photographic emulsions, 34, 122
Photosphere, the apparent surface of the Sun, 37, 38
Physiology of the human eye, 21
Piggyback mounting bracket, 27

Pleiades cluster, 131
Plotting the orbital position of a planet, 64
Polar alignment, 10, 11, 20
The polar axis of the Sun, 42
Polar axis has a drive, 9
Polaris relative to the celestial pole, 15, 17
Precession, 7
Printing the digital image of a variable star, 118
Processing the print, 33
Processing solar photographs, 42
Proper motion, 7, 139–141
The proper motion of Barnard's star, 139–141

Radial velocity, 139
RASC, 11, 42, 43, 60, 75, 76, 82, 83, 110
Reciprocity failure, 29
Relationship between the apparent and true orbits of a binary star, 104
Relative sunspot numbers, 42
Remote shutter control or a self-timer, 31
Resolving power of a telescope, 22
Reticle, 15, 27, 75, 92
The retrograde motion of Mars, 73, 74
Right ascension, 2, 3, 7, 8, 13, 14, 15, 17, 19, 64, 71, 74, 80, 91, 93, 113, 136
Rigid mounting, 9

Scaliger, 8
Scanner with 35-mm slide and negative scanning capability, 35
Seeing, 23, 25, 61, 64, 75, 78, 93, 114
Selenographic colongitude, 53
Selenographic longitudes, 53
Serendipitous comet discoveries, 82
Serious observing, 10, 51
Setting circles, 13, 17, 20, 83, 113
Short focal length refractor, 21, 35
Short focal length telescopes with low power eyepieces, 110
Short focus low power instrument best for

observing nebulae and star clusters not necessarily true, 123
Shutter speeds, 31, 52
Sidereal day, 7
Sidereal time, 7
Single lens reflex film cameras, 35
Size of the secondary mirror, 25
Sketching the surface of a planet seen through a telescope, 61
Small telescopes, 9, 13, 25, 28, 30, 34, 59, 92, 100, 125
Solar equator, 37
Solar latitude, 37, 42, 48, 49
Solar Observation Section of the AAVSO, 42
Solstice, 1–2
Spectrum of Sirius, 137
Speers–Waler eyepieces, 26
Spherical aberration, 23
Spherical rather than parabolic mirrors, 26
Stability, 9
Stellar spectra, 34, 133, 134, 135, 136
Stellar spectra on black-and-white film produces greater image contrast and resolution of absorption lines, 135
Stonyhurst disks, 43
Stonyhurst grids, 34
Sun's differential rotation, 49
Sundials, 6
Sunspot cycle, 37, 38
Sunspots, 37, 38, 39, 41, 42, 48
Sunspots vary considerably in size and structure, 39

Tabletop legs, 10, 19
Telephoto conversion lens, 32
Theophilus, 56
Time zones, 6
Tracking the positions of an asteroid, 83
Traditional method of viewing the Sun, 39

Universal Time, 8, 41, 43, 114

Variable star calibration scale, 118
Variable star data analysis, 120
Variable star magnitudes, 110, 118

Index

Variable stars, 25, 28, 29, 30, 31, 109–115, 117–122, 125
V-Block filter, 24
Visual binary stars, 91–101
Visual observations, 10, 13, 39, 41, 49, 51, 74, 80, 109–115, 122
Visual observations of comets, 80
Visual observations of Jupiter, 67
Visual observations of the Moon, 51
Visual observations of the Sun, 39
Visual observations of variable stars, 109–115
Visual record of the motion of Mars, 74

Which small telescope should you buy, 27
Wide field photographs, 33

Year solar magnetic polarity cycle, 37
A young star cluster, 130, 131

Zenith, 1
Zero-age main sequence, 131
Zodiac, 7
Zoom lens, 31
Zurich classification system, 40

Other Titles in this Series

(continued from page ii)

Lunar and Planetary Webcam
User's Guide
Martin Mobberley

Real Astronomy with Small Telescopes:
Step-by-Step Activities for Discovery
Michael K. Gainer

Human Vision and the Night Sky: How
to Improve Your Observing Skills
Michael Borgia

CCD Astrophotography: High Quality
Imaging from the Suburbs
Adam Stuart

Pattern Asterisms: A New Way
to Chart the Stars
John Chiravalle

Urban Astronomer's Guide
Rod Mollise

How to Photograph the Moon and Planets
with Your Digital Camera
Tony Buick

Digital Astrophotography: The State
of the Art
David Ratledge

Printed in the United States of America